WEST

W 5083344 8

WESTHILL COLLEGE OF EDUCATION LIBRARY

Birmingham, 29.

A fine will be charged if this book is returned after the last date stamped below

14 JUN 1974

27085

LUCAS, H.

Statistical methods

B. L. & C.

Book No. 27085

Author LUCAS, H

Title Statistical methods

519.9 LUC

AF849347

# Statistical Methods

Harold Lucas, B.Sc.(Econ.), B.Com., A.I.S., A.M.B.I.M., A.C.C.S.
Senior Lecturer at Barking Regional College of Technology

LONDON
BUTTERWORTHS
1970

*ENGLAND:* BUTTERWORTH & CO. (PUBLISHERS) LTD.
LONDON: 88 Kingsway, W.C.2.

*AUSTRALIA:* BUTTERWORTH & CO. (AUSTRALIA) LTD.
SYDNEY: 20 Loftus Street.
MELBOURNE: 343 Little Collins Street.
BRISBANE: 240 Queen Street.

*CANADA:* BUTTERWORTH & CO. (CANADA) LTD.
TORONTO: 14 Curity Avenue, 374.

*NEW ZEALAND:* BUTTERWORTH & CO. (NEW ZEALAND) LTD.
WELLINGTON: 49/51 Ballance Street.
AUCKLAND: 35 High Street.

*SOUTH AFRICA* BUTTERWORTH & CO. (SOUTH AFRICA) LTD.
DURBAN: 33/35 Beach Grove.

WESTHILL COLLEGE LIBRARY
27085
18 · 2 · 71
T1980 - £1 · 50
519 · 9

©
Butterworth & Co. (Publishers) Ltd.
1970

ISBN: O 406 75580 9

*Printed in Great Britain at*
*the St. Ann's Press, Park Road, Altrincham*

# Preface

This text-book is designed for the increasing numbers of students and others in the fields of management, business and the social sciences who find a knowledge of statistical methods is essential to their studies, or is necessary for the effective performance of their work.

Management students will find the normal requirements of courses such as those leading to the Diploma in Management Studies are dealt with in a way that relates the subject to their many other topics of study.

Business students taking O.N.C. or O.N.D. and continuing through to H.N.C. or H.N.D. should find all their needs for statistical methods are met. Many business applications are outlined and a link is provided with the use of statistics in surveys and experiments as well as a brief account of published statistics. The student who wishes rapidly to catch up with the lower level work will find the earlier chapters can quickly be assimilated.

Many professional students require a basic knowledge of statistics at some stage of their studies whether or not this forms an actual examination subject. Examples have been drawn from many areas of business and social activity including, for example, accounting and costing, marketing, personnel management, and economics and the other social sciences. The requirements of bodies such as the Institute of Cost and Works Accountants should be met, and chapter 21, which is intended as a mathematical refresher, seems to cover, in brief outline, the new business mathematics syllabus which that body will run in conjunction with its statistics examination, though it is not intended as a text-book on business mathematics.

Students of economics and the social sciences should find that their needs for a background of statistical techniques are satisfied. Many of the C.N.A.A. degree schemes in such subjects have requirements similar to the coverage offered.

Statistics is basically a mathematical subject but few of such readers will be expert mathematicians. This has been taken into account in the presentation.

The book is built around practical examples with logical explanation. Rigid mathematical proofs are not given and are not required by the syllabuses to which students will normally be working. At the same time the reader with more than average mathematical knowledge should find that the subject matter is of interest and value and is presented quite differently from the normal mathematical text. The danger of a more abstract approach is that attention may focus on the mathematics rather than on the practical features of the problem being handled. The intention is to train people in the uses of statistics rather than to produce statisticians.

The chapters build up in logical sequence. Some teachers may prefer to deal with chapter 6 (dispersion) immediately after chapter 3 (means). If, however, a course on published statistics or on surveys is being run parallel with the statistical methods students will wish to be acquainted with index numbers (chapter 5) at an early stage. More advanced material has been collected in chapter 20. With the gradual raising of standards most of this will probably be expected of students in the near future and does now feature in many courses.

Subjects of fundamental importance have been treated at some length, examples being averages and dispersion. Such topics normally feature prominently in examinations, either as questions in their own right or as essential bases of calculations for more advanced questions. Emphasis has been placed on the interpretation of data, and the various parameters are regarded as serving a purpose, rather than as being exercises in calculation.

An attempt has been made to stress the probabilistic basis of statistics, without the wholesale sacrifice of much that was of value in the more traditional approach to this subject for those who do not intend to be statisticians in the professional sense.

At the end of many of the chapters will be found selected questions with model answers, chosen to supplement the text. *Here square brackets enclose comments which would not appear in the answers.*

Typical questions are included to afford practice to the reader.

Published statistics have been treated briefly (chapter 18). The information can change rapidly and reference to up-to-date government and other journals affords a good means of keeping abreast, as well as being an essential part of any realistic scheme of studies.

I am indebted to the Literary Executor of the late Sir Ronald A. Fisher, F.R.S., to Dr Frank Yates, F.R.S., and to Oliver & Boyd Ltd., Edinburgh, for permission to reprint Table 4 from their book *Statistical Tables for Biological, Agricultural and Medical Research.*

# Contents

# 1

# Introduction to Statistics

## 1.1 THE SCOPE OF STATISTICS

*Statistics*, in the singular, is the body of theory and methods concerned with the *collection*, *processing*, *analysis*, *interpretation* and *presentation* of data.

In a world increasingly dependent on information statistics plays a vital role. Techniques are provided which are of especial importance in business and management and in the social sciences. These fields of activity demand both a comprehensive view of a community, or of a market, or of a business, and also a more detailed study of the components and their interrelationships. Modern administration, whether in the public or private sector, relies on data to provide a factual basis for decisions. The subject we are studying attempts to provide guidance in acquiring and dealing with such data.

The facts and figures which provide the raw material of our study are generally referred to as *statistics*, in the plural. A characteristic of statistics is that they are plural, implying that there are a number of items. They refer to groups of items or, perhaps, to a series of data over a period of time. An individual item may be considered but it is of statistical interest only in so far as it is compared with other items which have facts in common with it.

*Descriptive* statistics are the most obvious to the man in the street. The best-known examples are those provided by government bodies, often published at regular intervals. Such data are normally intended for use by a wide variety of persons in fields such as business, economics and sociology. The term *secondary data* is often used to distinguish such statistics from data especially collected for a specific investigation and referred to as *primary data*. Alternatively the terms primary and secondary can be used to describe data as originally collected (raw data) and data after treatment by statistical methods (processed data), respectively.

The collection of statistics relating to the numbers and conditions of populations goes back to ancient times, though the modern form of *census* started in the nineteenth century. This term is now extended to cover surveys of topics such as distribution and production in a community, as well as purely population studies. Terminology does not seem to be generally agreed. We speak of *surveys* to imply a comprehensive appraisal of the characteristics of a situation. This term is, however, often restricted to imply a sampling approach, in which case the word census generally implies an attempt to take all members of the population into account.

Statistical methods help in the devising of censuses and surveys and in the collection and presentation of such data. At first sight this seems a simple matter but it is easy to waste money and obtain unreliable results. *Sampling* is often used and forms one of the basic tools of statistics.

The uses of such general data to administrators at national and local level and also to business men in connection with activities such as marketing and personnel are not hard to see.

Many data also arise within individual business units. These statistics may relate to finance and costs, production, marketing, personnel, development and to almost anything of concern to the management. The techniques may range from calculating the average output per man-hour or the ratio of profit to sales up to sophisticated attempts at business forecasting and the control of production processes.

Statistics are used for many purposes. The historian uses them to clarify the past. The manager is interested in this only in so far as it may provide a guide to action in the future. In this connection the aspects of statistics devoted to the analysis and interpretation of data are important.

*Variability* is a characteristic of the data handled by statistical methods. It implies that there will be differences between the items being studied. Even though the items may be summarised in an average the spread about this average can be of great concern.

In marketing it is of use to have information on the average sales achieved through the branches of an organisation. It is much more revealing if the variability between these branches can be considered without having to refer to all the basic data. If we can further assess the extent to which this variability is changing from period to period we may have a result of great use in planning.

In psychology a knowledge of the average intelligence of a group is of much greater value if we know how this stands in relation to the rest of the population and, also, if we have an indication of the spread within the group itself.

On the production side of industry the measurements of successive items produced on an automatic machine will differ from each other. The degree of precision in measuring determines how obvious these differences are, but they will be found to cluster round an average according to a law which can be stated precisely. This can be used to control production, implying that we know the characteristics of what is being produced and that the variability is within the limits allowed in our plan.

Variations which occur over periods of time can be studied, such as sales of a firm, trade union membership or the balance of payments. Business and national forecasting makes use of such *time series.*

Again we have the study of how, if at all, variations in one factor are related to those in another. This is the area of *regression* and *correlation,* of which the statistical link between cigarette smoking and lung cancer is probably the best-known example.

Many of our statistical techniques originated in the variability inherent in life, whether human, animal or plant, and in the need to make comparisons.

*Probability* enters into much statistical work. Foundations were laid in the eighteenth century in connection with gambling and the study of the laws of chance. It is now recognised that such knowledge is basic to modern business in so far as this is concerned with trying to anticipate the future and making decisions in uncertain situations.

The degree of uncertainty involved when we are assessing data is really a statement of our lack of knowledge or of the extent to which random elements may influence the outcome.

When items are dealt with individually the element of uncertainty is often great, becoming less as more items are considered. Assurance offers a good example. Rarely would we be in a position to say that a selected individual will live for a stated length of time. As far as all people of that age and sex and country are concerned, however, we may make this statement in general terms.

Planning is often concerned with the outcome over several events termed the *expectation*. At a level more in contact with individuals the variability between these becomes more significant.

The failure of lamps in an office lighting system will follow a predictable pattern but we cannot state precisely when any individual lamp is going to fail. The lamps have an average life and the probability of any lamp varying from this by a stated amount can be assessed. We can also state how many lamps will probably have failed, out of a large number, after any selected period of time.

In the field of personnel we can assess the overall results of using a selection test and decide whether it is worth the expense involved. This turns on the average improvement in candidates selected using the test compared with those selected without using it. We cannot, however, guarantee that the selection of any particular individual will be correct.

The concept of the *norm* (or *average*) tends, in the social sciences, to replace the precise statements of the physical sciences. Even in subjects as precise as physics, however, it is a matter of the level at which we are studying. There is a precise relationship between the volume of a gas and its absolute temperature, but when the physicist deals with individual molecules in conditions of temperature change he cannot make precise statements but has to talk in terms of probability.

Management is becoming more scientific. *Operational research* attempts to place mathematics at the disposal of managers in a practical way. The problem to be studied is first decided. A mathematical *model* of the actual or proposed situation is then constructed. This may be just a simple equation in the more elementary cases, but it often implies taking into account the variability and probability aspects of a real-life situation. Here statistical methods are of great use, especially *simulation* techniques, which allow the random element to be represented. After a satisfactory model has been constructed and tested it is then operated to see what happens in various circumstances. The outcome is noted and translated back from mathematical into day-to-day terms.

The use of statistical and operational research methods has been enormously assisted by the development of digital *computers* which enable complex data to be handled at a speed and accuracy previously unthinkable.

## 1.2 PRACTICE QUESTIONS

It is optimistic to expect that the newcomer to statistics will be able to give anything like adequate answers to questions relating to the general application of statistical techniques after reading only a few pages. The following questions would probably be better answered at the end of a statistics course. Throughout his studies the student should be trying to see practical applications of statistical methods in his work and in his other studies and should read accounts of modern applications such as are to be found in professional journals.

1. Explain how a statistical approach to problems may be of help to the businessman, drawing on at least three different aspects of business activity.

2. Economists, sociologists and students of management and business have been trained in statistical methods as a matter of course whereas statistics has only recently come to be included in some engineering and science courses. Try to account for this.

3. Statistics has been called the science of counting. Discuss this definition critically.

4. Give examples of the way in which statistics can be of help in either production management or personnel management or marketing or industrial research and development.

5. 'Econometrics is really applied statistics.' Do you agree. Support your views.

6. 'It is impossible to imagine the development of sociology without statistics.' Comment critically.

7. 'The statistician takes simple facts and complicates them.' Discuss, showing your appreciation of the proper role of statistics.

# 2

# The Pattern of Data

## 2.1 INTRODUCTION

One of the basic aims of statistics is to organise *raw data* in such a way that any underlying pattern is revealed.

Raw data may be regarded as the figures arising originally from an investigation. No order or system may be apparent. If the data are numerous it may be impossible easily to appreciate the salient points. The average person will feel a psychological resistance to such data, while the busy manager will not have the time to sort them out.

This chapter is concerned with some of the elementary ways of presenting data. These methods have the object of making data interesting and acceptable. At the same time they point the way to a more mathematical approach to statistics.

Table 2.1 is an example of raw data. The reader may find it useful to examine the figures to see what they convey to him or her before reading further.

Table 2.1. *Bonus data for 150 employees*

| a | b | c | d | e | f |
|---|---|---|---|---|---|
| 5·30 | 4·00 | 6·15 | 4·25 | 5·55 | 5·45 |
| 3·05 | 5·90 | 3·85 | 5·75 | 5·75 | 4·00 |
| 4·75 | 6·50 | 5·75 | 6·10 | 4·25 | 6·60 |
| 3·60 | 3·45 | 4·25 | 6·00 | 4·50 | 7·55 |
| 6·10 | 4·70 | 4·65 | 5·85 | 4·50 | 4·60 |
| 3·75 | 5·30 | 5·75 | 5·00 | 4·50 | 6·40 |
| 4·10 | 5·75 | 6·00 | 5·00 | 4·75 | 6·25 |
| 6·25 | 5·75 | 6·00 | 5·00 | 5·40 | 6·90 |
| 4·15 | 5·75 | 6·05 | 5·60 | 5·40 | 6·50 |
| 5·80 | 5·75 | 5·75 | 6·00 | 6·90 | 5·85 |
| 4·25 | 5·75 | 6·00 | 5·25 | 6·75 | 5·25 |
| 5·05 | 5·80 | 5·20 | 5·25 | 5·65 | 6·30 |
| 5·10 | 5·75 | 5·25 | 6·00 | 5·50 | 4·75 |
| 4·45 | 4·00 | 6·10 | 5·25 | 4·90 | 6·25 |
| 5·60 | 4·55 | 5·00 | 5·25 | 5·50 | 5·85 |
| 4·50 | 5·75 | 5·60 | 5·65 | 4·95 | 5·95 |
| 4·50 | 5·40 | 5·00 | 5·50 | 5·50 | 4·45 |
| 4·50 | 6·65 | 5·25 | 4·80 | 5·90 | 4·20 |
| 6·00 | 5·75 | 5·50 | 5·50 | 4·45 | 6·30 |
| 6·55 | 6·75 | 5·00 | 4·90 | 5·95 | 5·45 |
| 5·65 | 6·45 | 5·60 | 4·95 | 4·20 | 5·45 |
| 5·65 | 4·60 | 6·40 | 5·00 | 5·95 | 5·50 |
| 5·65 | 5·45 | 6·00 | 5·25 | 5·50 | 4·40 |
| 3·35 | 6·50 | 5·25 | 5·90 | 5·90 | 5·45 |
| 4·80 | 5·75 | 5·70 | 6·20 | 6·00 | 5·45 |

The title is not very revealing. In practice we should require much additional information such as the name of the firm, the types of employees (staff or manual?), the date and the period of time involved (week or month?), the order in which the items appear (alphabetical or random?) and the units in which the items are expressed (pounds or dollars?). Such details should normally be included in the titles of tables and diagrams, perhaps in the form of footnotes. Often, in this book, an abbreviated title will be used, since the data serve primarily to illustrate statistical techniques.

As for the detail of the table we should probably consider factors such as the number of items, the largest and smallest items, how items differ and how they are spread out, and whether there seems to be a representative item.

Statistics approaches these matters in a systematic way.

## 2.2 ARRAYS

The most obvious approach to making data more amenable and revealing is to put them in order or to *rank* them. This has been done in table 2.2. Such a table, in which all the data are included, moving from the smallest to the largest (or vice versa), is called an *array*.

Table 2.2. *Array of bonus data (partial)*

| | | |
|---|---|---|
| 3·05 | 5·45 | 6·20 |
| 3·35 | 5·45 | 6·25 |
| 3·45 | 5·45 | 6·25 |
| 3·60 | 5·45 | 6·25 |
| 3·75 | 5·45 | 6·30 |
| 3·85 | 5·45 | 6·30 |
| 4·00 | 5·50 | 6·40 |
| 4·00 | 5·50 | 6·40 |
| 4·00 | 5·50 | 6·45 |
| 4·10 | 5·50 | 6·50 |
| 4·15 | 5·50 | 6·50 |
| 4·20 | 5·50 | 6·50 |
| 4·20 | 5·50 | 6·55 |
| 4·25 | 5·50 | 6·60 |
| 4·25 | 5·55 | 6·65 |
| 4·25 | 5·60 | 6·75 |
| 4·25 | 5·60 | 6·75 |
| 4·40 | 5·60 | 6·90 |
| 4·45 | 5·60 | 6·90 |
| 4·45 | 5·65 | 7·55 |
| next 45 items omitted | next 45 items omitted | |

It is quite a labour to derive an array from an original table of random data. A simple approach is to tick off each item from the original table, at the same time copying it roughly in position on a new sheet, leaving plenty of room for intermediate items as they occur. A useful way with a large number of items is to write each on a separate sheet or card so that they can be assembled in order. This can be helpful if there is more than one fact to be recorded about each item, since the cards can be arranged in different ways. If the data can be put on to punched cards the problem of arraying is solved by passing the cards through the sorter which will put them in order ready for the automatic printing of the array. A computer can also be programmed to array data.

## 2.3 GROUPED ARRAYS

A major disadvantage of the ordinary array is that it does not compress the data. We are left with a table the same length as the original.

This may be overcome by a *grouped array* as in table 2.3. The 150 items have been divided into groups or *strata* each containing one-tenth of the items in order. The first stratum has the first 15 items, the second the next 15, and so on. The size of the smallest and largest item in each stratum is shown. Any other division of the data such as fifths, quarters or hundredths may be used to suit particular circumstances.

Table 2.3. *Grouped array of bonus data*

| Stratum | from | to |
|---|---|---|
| 1 | 3·05 | 4·25 |
| 2 | 4·25 | 4·60 |
| 3 | 4·65 | 5·00 |
| 4 | 5·00 | 5·25 |
| 5 | 5·30 | 5·50 |
| 6 | 5·50 | 5·70 |
| 7 | 5·75 | 5·80 |
| 8 | 5·80 | 6·00 |
| 9 | 6·00 | 6·30 |
| 10 | 6·30 | 7·55 |

The chief use of the grouped array, or of measures derived from it, is probably in connection with social statistics. We often think in terms of a stratified population. Percentile rankings, showing where a person would fit into a sample of 100 people drawn from the general population, are increasingly used for the assessment of intelligence and other personal factors.

We shall see in the next chapter that the values at which the data are split into tenths, quarters etc. have a special name. They are called *partition values*. Most frequently they have to be derived from data which has been compressed and where the full array is not available.

## 2.4.1 FREQUENCY DISTRIBUTIONS

This is probably the most important way of tabulating data. It is the starting point for much of the analytical work which forms the bulk of statistical method.

The first step in preparing a frequency distribution is to decide the appropriate groups into which to divide the items.

It is useful first to determine the range of the items in round numbers. In the case of the bonus data the items range roughly from £3 to £8.

Next decide the width of the groups which are to be used and also the number of groups. One decides the other. Too narrow groupings result in a large number of groups, making it difficult to appreciate the pattern, and sometimes giving groups of zero frequency. Groups which are too broad obscure the detail of the pattern. Much depends on the total number of items. The larger this happens to be the narrower can be the groupings without irregularities appearing.

*Sturges' rule* provides a formula for the number of groups (k) desirable when there are N items in total:

$$k = 1 + 3{\cdot}3 \log_{10} N.$$

For the 150 bonus items this gives 8 groups. Practical considerations, such as the need to have an acceptable group width, may modify this.

The problem of allocating items to their respective groups has to be considered. If the group runs from 4 to 4·5 how is this interpreted? Is an item exactly 4·5 included in this group or in the group 4·5 to 5? It is best to make this quite clear by showing the group as *from 4 to 4·9* or as *4 and less than 4·5*. If this is not done the allocation must be consistent. The most usual method would be to allot an item on the border to the higher group. If many items occur at the margins of the groups it would be well to re-arrange the grouping so that such items clearly fall within particular groups, preferably near the group centres. An approach sometimes used is to allocate marginal items alternately to upper and lower groups. Much trouble can often be saved by a careful examination of the data before the precise grouping is decided.

When the frequency distribution has been prepared it tends to be regarded subsequently as the basic data for calculations. When we look at a group designated *4 to 5* we think of the items in that group as being centred on 4·5, and many calculations will also be based on that assumption.

Generally group widths should be kept the same throughout the table. An exception may occur in the case of wider groups at the beginning and end used to gather together scanty items. Groups displaying points of special interest may sometimes be shown on a narrower grouping. If unequal groupings are used care must be taken in drawing diagrams to illustrate the data and in performing calculations, as is explained later.

If an array is available the task of deriving a frequency distribution is greatly simplified. We just read down the array counting the number of items from one group boundary to the next.

When raw data has to be used the *gate method* can speed the job and reduce the risk of error. A blank table is prepared with the groups indicated in order on the left. As each item is allocated to its group it is ticked in the original table and a stroke is made vertically in the blank table opposite the appropriate group. Every fifth stroke is made across the previous four so that a gate is drawn. This facilitates reckoning the final total of each group, as each gate is counted as 5 and any odd strokes are added to this. Thus ~~////~~ ~~////~~ ~~////~~ // is at once recognised as 17.

Always check that the total of the group frequencies is the same as the total of items in the original data.

The effect of using groups of different widths is shown in tables 2.4 and 2.5. In the first of these the group widths tend to be narrow and the table is rather lengthy. The second table gives a grouping about as broad as possible without obscuring the pattern.

An alternative to the term *group width* is *class interval*. Another point of terminology is that figures read vertically are termed *columns* while those read horizontally are referred to as *rows*.

Table 2.4. *Frequency distribution of bonus data (based on 0·5 grouping)*

| Bonus | Frequency |
|---|---|
| 3 and less than 3·5 | 3 |
| 3·5—4·0 | 3 |
| 4·0—4·5 | 15 |
| 4·5—5·0 | 20 |
| 5·0—5·5 | 30 |
| 5·5—6·0 | 45 |
| 6·0—6·5 | 23 |
| 6·5—7·0 | 10 |
| 7·0—7·5 | — |
| 7·5—8·0 | 1 |
| Total | 150 |

Table 2.5. *Frequency distribution of bonus data (based on unit group width)*

| Bonus | Frequency |
|---|---|
| 3—4 | 6 |
| 4—5 | 35 |
| 5—6 | 75 |
| 6—7 | 33 |
| 7—8 | 1 |
| Total | 150 |

## 2.4.2 PERCENTAGE FREQUENCIES AND PROBABILITIES

It is sometimes desirable to express the frequencies as percentages of the total frequency. Each frequency is divided by the total frequency and multiplied by 100 to give the required figure. This has been done in table 2.6.

Table 2.6. *Percentage frequencies of bonus data*

| Bonus | % Frequency |
|---|---|
| 3—4 | 4·0 |
| 4—5 | 23·3 |
| 5—6 | 50·0 |
| 6—7 | 22·0 |
| 7—8 | 0·7 |
| | 100·0 |

Such presentation is a simple way of *standardising* the frequencies of data. We may wish to compare two or more sets of workers with quite different totals in each set. The use of percentages has the effect of making the totals for comparison the same, so that attention is focused on any internal differences.

Instead of using percentages we may divide each frequency by the total frequency without multiplying by 100. This results in the sum of the frequencies of the groups adding up to one. It is useful if we are thinking in terms of *probabilities*, since total probability, or certainty, is usually represented by unity. Thus, if we take the 150 bonus earners, an individual is bound to occur somewhere in the table. The probability that a person taken at random received between £6 and £7 is 0·22 (i.e. 22% or slightly more than one chance in five). This aspect of frequency distributions is particularly important for the more advanced treatment of data when we are concerned not with just the pattern of the data but with its implications.

## 2.5.1 HISTOGRAMS

Frequency distributions may be illustrated by means of *histograms.*

Diagram 2.1 is a histogram of the bonus data shown in table 2.5. It will be seen that vertical columns are drawn to indicate the frequency of each group, shown on the horizontal axis.

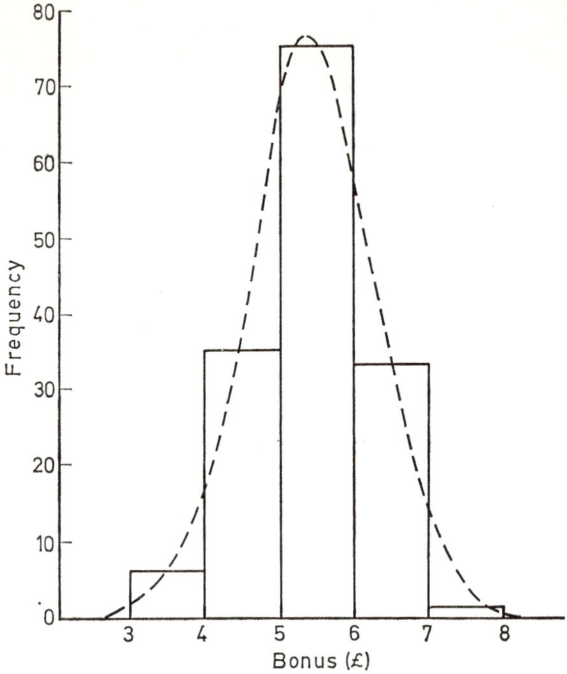


Diagram 2.1. *Histogram of bonus data,* with smoothed histogram shown by broken line

Histograms always show the frequency of occurrence of some variable. It is usual to show the frequencies vertically but occasional examples are to be found where they are shown horizontally. This is often the case with age distributions where the ages are shown vertically with the number of men in each age group horizontally on one side, and the number of women on the other (see diagram 16.3).

The histogram should be distinguished from the *historigram.* The latter is concerned with the *history* of some data, for example how the sales of a business have varied over the years. In this case the horizontal axis denotes time, not a variable such as bonus data.

Strictly the term *histogram* should perhaps be reserved for a diagram of the probability type, where the frequencies have been made to total unity. Generally it has come to be used in connection with any diagram similar to 2.1.

It must be stressed that it is the *area* that is important with histograms and the diagrams derived from them. The frequencies are proportional to areas. With groups of equal width this implies that the frequencies are also proportional to the heights of the vertical columns.

## 2.5.2 EFFECT OF UNEVEN GROUPING

It is well to draw histograms with groups of equal width so that the eye can make its judgment from the heights of the columns alone.

If groups of varying width are used the visual impression conveyed is misleading unless the frequencies are adjusted. For example, if the 5—6 group in table 2.5 were divided into two groups 5—5·5 and 5·5—6 the frequencies of these groups would be 30 and 45 respectively. If the 6—7 and the 7—8 groups were combined the frequency would be 34. The effect of plotting these frequencies is shown by the broken line in diagram 2.2. The impression conveyed is quite different from that when the items are evenly grouped.

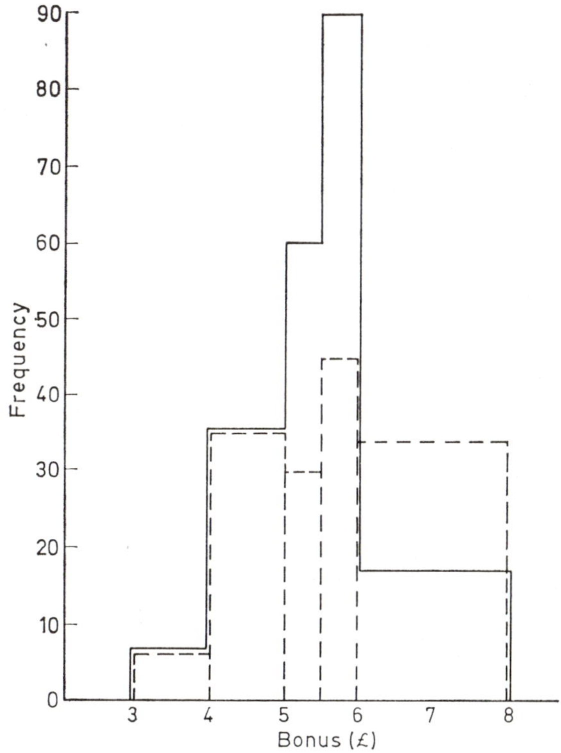


Diagram 2.2. *Histogram with uneven group widths.* The broken line shows the misleading impression which can be corrected by adjustment of frequencies, as shown by the full line

The way to redress the visual balance is to multiply the narrower group frequencies by 2 to allow for the fact that their width is only half that of the normal group. This gives artificial frequencies of 60 and 90 respectively, of which the average is the original 75. The wider group frequency is divided by 2 to give artificial frequency of 17.

The solid line in the diagram shows that the visual effect of these adjustments is similar to that of the evenly grouped data.

The rule for adjustment is

$$\text{adjusted frequency} = \text{actual frequency} \times \frac{\text{normal group width}}{\text{special group width.}}$$

The fact that the frequencies have been adjusted should be stated in a note to the diagram. It is not desirable to have a diagram with variations in group width since this makes comprehension difficult.

Many items of interest to businessmen and economists suffer from this problem of grouping and are best presented in some other form of diagram, such as the Lorenz curve (see section 6.10). Examples of such data will be found in tables 3.10 and 3.11, relating to union membership and the size of manufacturing establishments.

### 2.5.3 SMOOTHED HISTOGRAMS AND DENSITY FUNCTIONS

The broken line of diagram 2.1 is an attempt to smooth the rectangular forms of the histogram into a continuous curve. This has been done by guess and different people would not draw precisely the same shape.

Many distributions that occur in practice are nearly the same as smooth curves for which we can derive the equations. This is the real significance of the title of this chapter. There are relatively few standard patterns needed to cover the vast bulk of practical data.

Such curves are often termed *frequency functions* or *probability density functions* (if the emphasis is on probability). Probability is dealt with more fully elsewhere in this book (especially in chapters 7 and 8). For the moment we may regard the area under a histogram as an indicator of probability. The whole area represents total probability, or certainty. This is equivalent to 100% or to 1, if we adopt the usual convention of probability theory. The area of any vertical slice, expressed as a fraction of the total area, represents the probability of events occurring from the lower limit of the slice to the upper limit, wherever these limits have been chosen.

The properties of some of these standard curves have been explored in depth and the results can be applied to the particular cases which concern us. Before we can understand them properly we have to learn something of the statistical measures on which they are based.

## 2.6 FREQUENCY POLYGONS

This is really a special form of histogram. The group frequencies are plotted above the centres of the groups to which they relate and the points are connected by straight lines, as in diagram 2.3.

It will be seen that the diagram is brought down to the base line at each extremity by connecting to the centres of artificial groups of zero frequency.

This treatment results in a diagram with an area identical to that of the original histogram. It can be seen from the inset in diagram 2.3 that the area

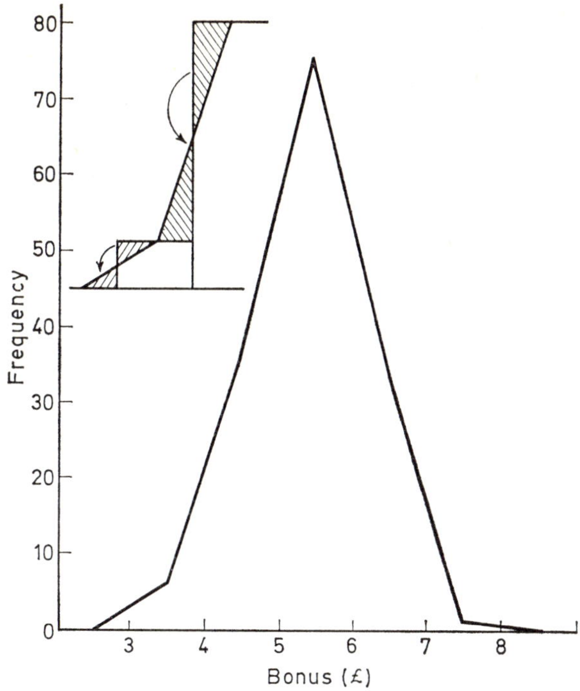


Diagram 2.3. *Frequency polygon of bonus data.* Inset shows how area cut off the histogram equals the area added to the polygon

cut from the corner of each rectangle is replaced by a similar area forming part of the polygon.

If the total frequency of items is large then we may consider the histogram as split into a great number of thin vertical columns and the frequency polygon will approximate to a smoothed histogram.

## 2.7.1 CUMULATIVE DATA

The *cumulative frequencies* may be obtained from a frequency distribution by successive addition of the group frequencies.

This has been done in table 2.7. It will be seen that both the original and the percentage frequencies have been treated in this way.

Table 2.7. *Cumulative frequencies of bonus data*

| Bonus | Frequency | Cumulative frequency ascending | Cumulative frequency descending | % Cumulative frequency ascending | % Cumulative frequency descending |
|---|---|---|---|---|---|
| *a* | *b* | *c* | *d* | *e* | *f* |
| 3—4 | 6 | 6 | 150 | 4·0 | 100·0 |
| 4—5 | 35 | 41 | 144 | 27·3 | 96·0 |
| 5—6 | 75 | 116 | 109 | 77·3 | 72·7 |
| 6—7 | 33 | 149 | 34 | 99·3 | 22·7 |
| 7—8 | 1 | 150 | 1 | 100·0 | 0·7 |
| | 150 | | | | |

When we accumulate the frequencies from the smallest item first we refer to the cumulative frequencies as *ascending*. If we start with the largest we say they are *descending*. The former is shown in columns *c* and *e*, and the latter in columns *d* and *f*. Normally we should select one or the other.

The interpretation of the cumulative data is important. The ascending figures may be read as:

6 people received up to £4,
41 people received up to £5, etc.;

and the descending figures as:

150 people received more than £3,
144 people received more than £4, etc.

An obvious check for calculations of this kind is that the final figure should agree with the column total of the group frequencies.

## 2.7.2 OGIVES AND DISTRIBUTION FUNCTIONS

The *ogive* is the graph of cumulative frequency data with the cumulative frequencies shown vertically.

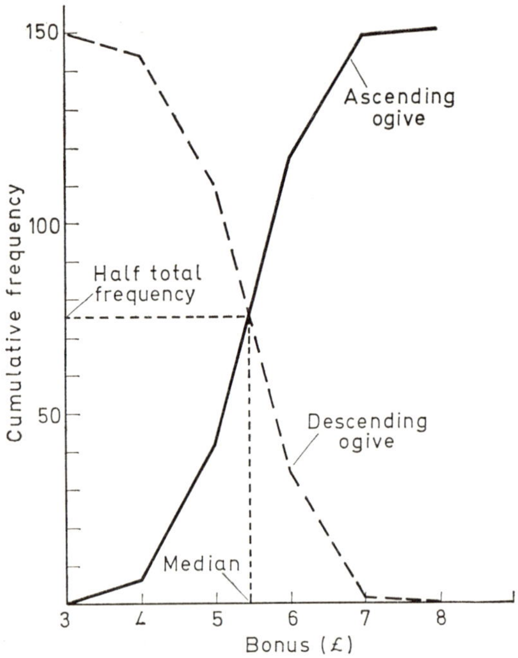


Diagram 2.4. *Ogives*, ascending and descending, of the bonus data, showing the median at intersection

Diagram 2.4 gives both the ascending and descending ogives for the bonus data. Normally only one of these would be used, generally the ascending version.

It will be seen that the cumulative frequencies are plotted over the *upper or the lower limits* of the groups, depending on whether the ascending or descending version is being drawn. This follows from the interpretation of cumulative data mentioned in section 2.7.1. A clear distinction should be made between this case and that of the histogram or frequency polygon, where the vertical columns are drawn over the widths of the groups, and the points of the polygon are plotted over the centres of the groups, respectively.

Estimates of the number of items occurring up to any chosen value on the horizontal scale can be obtained from the ogive. A vertical line is drawn at the chosen value and the cumulative frequency at which it cuts the curve is read. The number of items between any two values can be found by repeating this for the second value and subtracting the frequency of the lower value from that of the higher.

The term *distribution function* is often used when the probability aspects of the ogive are being considered. In this case the total frequency will be unity.

Distribution functions and density functions (2.5.3) are obviously related. The former shows the area under the latter up to any given point. Conversely the rate at which the distribution function is increasing at any point, measured by it gradient, is indicated by the frequency (or probability) of the density function at that point. Readers with a knowledge of calculus will appreciate that this is a matter of integration or of differentiation, depending on where we start. Simple calculus is explained at 21.5 but is not called for in many syllabuses.

## 2.8 WORKED EXAMPLE

*Question*

*a.* Draw a diagram to illustrate the age data (table 2.8) bringing out clearly differences in the age structure at the two dates.

*b.* Show the two age structures by means of ogives on the same base and briefly comment on their shapes.

*c.* From the ogive of the 1961 data estimate the number of people between the ages of 25 and 45.

*d.* Comment on possible inaccuracies in this estimate.

Table 2.8. *Age distribution of home population of U.K. census figures (thousands)*

| Age | 1901 | 1961 |
|---|---|---|
| 0— 9 | 8,487 | 8,028 |
| 10—19 | 7,762 | 8,003 |
| 20—29 | 6,981 | 6,563 |
| 30—39 | 5,328 | 7,083 |
| 40—49 | 4,002 | 7,100 |
| 50—59 | 2,801 | 6,958 |
| 60—69 | 1,808 | 5,019 |
| 70—79 | 847 | 2,937 |
| 80 and over | 219 | 1,017 |
| Totals | 38,235 | 52,708 |

(*Registrars General*)

*Answer*

*a.* [Since the total populations differ at the two dates it is necessary first to convert the two columns into percentages of their respective totals. This will emphasise the differences in composition of the two sets of data.]

Table 2.9. *Percentage age distribution of U.K. home population*

| Age | 1901 | | 1961 | |
|---|---|---|---|---|
| | % | Cum. % | % | Cum. % |
| 0— 9 | 22·2 | 22·2 | 15·2 | 15·2 |
| 10—19 | 20·3 | 42·5 | 15·2 | 30·4 |
| 20—29 | 18·3 | 60·8 | 12·5 | 42·8 |
| 30—39 | 13·9 | 74·7 | 13·4 | 56·3 |
| 40—49 | 10·5 | 85·2 | 13·5 | 69·8 |
| 50—59 | 7·3 | 92·5 | 13·2 | 83·0 |
| 60—69 | 4·7 | 97·2 | 9·5 | 92·5 |
| 70—79 | 2·2 | 99·4 | 5·6 | 98·1 |
| 80 and over | 0·6 | 100·0 | 1·9 | 100·0 |
| | 100·0 | | 100·0 | |

[Table 2.9 shows this conversion which could be done on a slide rule. Logarithms could also be used. With approximations of this kind it is possible for the total not to come precisely to 100·0.]

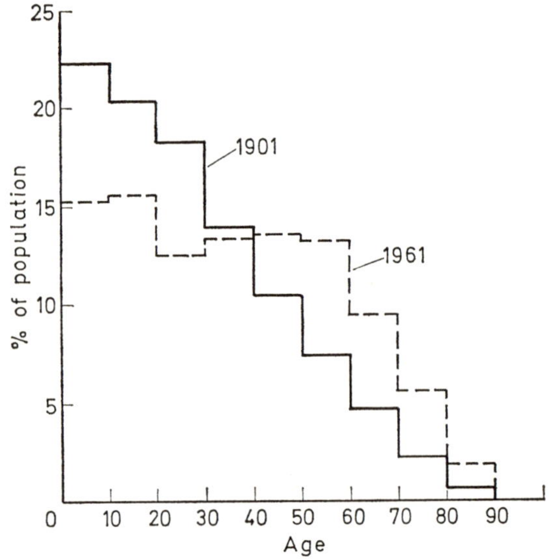


Diagram 2.5. *Histograms for age distributions,* 1901 and 1961

[Diagram 2.5 shows two histograms illustrating the percentage figures and treated in a way to bring out the differences in percentage composition.]

*b.* [For the ogives we need the cumulative percentage figures, which are also shown in table 2.9. Diagram 2.6 gives the required ogives.]

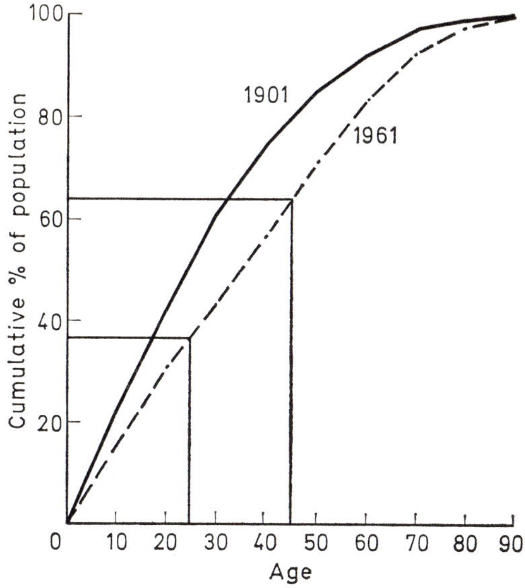


Diagram 2.6. *Ogives for age distributions,* showing estimation of frequencies

The gradient of an ogive at any point is an indication of how quickly the total at that point is increasing as we go from group to group.

The 1901 ogive has a steeper gradient than that for 1961 up to about the age of 30. After the age of about 40 its gradient is less than the 1961 line. This indicates a preponderance in the younger age groups in 1901.

*c.* [The number of people between the ages of 25 and 45 can be estimated from the ogive by drawing vertical lines at the points where ages 25 and 45 occur and reading off horizontally the percentage figures. These can be converted to actual numbers and the required figure found by subtraction. The subtraction can, alternatively, be done with the percentages and the difference immediately converted.]

| | % | | *Number* |
|---|---|---|---|
| Up to 45 years | 64 | $64 \times \frac{52 \cdot 7}{100} =$ | 33·7 |
| Up to 25 years | 36·5 | $36 \cdot 5 \times \frac{52 \cdot 7}{100} =$ | 19·2 |
| Difference | 27·5 | | 14·5 |

Check $27 \cdot 5 \times \frac{52 \cdot 7}{100} = 14 \cdot 5$

The estimated population between the ages of 25 and 45=14·5 million.

*d.* This method is liable to two sources of inaccuracy:
   i. That inherent in estimating from a graph. The values of 64 and 36·5 are really just approximations. A more accurate result can be obtained by calculation [see section 3.3.3].
   ii. That inherent in the assumption that the items are spread evenly throughout each group. We imply this by connecting the points by a straight line.

[N.B. Reference to more detailed information than is available in the question shows that the actual figures are 26·2% representing a population of 13·8 million. The items are not evenly spread being respectively 3,305, 3,258, 3,403, 3,680 and 3,442 for the 5 year groups from 20 to 44 years.]

## 2.9 PRACTICE QUESTIONS

1. Draw a histogram to illustrate the data of table 3.3.

2. Illustrate the data of table 3.4 by means of three frequency polygons on the same base. Comment on the resulting diagram.

3. Take alternate items of table 2.1 (giving 75) and put them into a suitable frequency distribution. Also prepare a frequency distribution for the remaining 75 items, and compare the two distributions.

4. Compare the following pairs of items:
   *a.* Histogram and ogive.
   *b.* Array and frequency distribution.
   *c.* Smoothed histogram and frequency polygon.

5. Take four of the six items in question 4 and give practical examples, from your particular field of studies, of the types of data for which they might be appropriate.

6. Draw a descending ogive for the data of table 2.8 (1961) and from it estimate the number of people up to 44 years old.

7. Form the 84 quarterly sales figures of table 14.1 into a frequency distribution and illustrate this by means of suitable diagrams.

As a practical exercise students should obtain frequency data by their own observation. Examples of topics would be: number of hours spent viewing television per week by individuals; the heights or weights of people; amounts of money spent in a particular way; and distances travelled each week.

# 3

# Averages

## 3.1 INTRODUCTION

An *average* attempts to compress data so that one figure can be regarded as representative of all items.

The chief averages, roughly in order of importance, are:

a. Arithmetic average or mean
b. Median
c. Mode
d. Geometric mean
e. Harmonic mean.

The choice of a suitable average in particular circumstances is a matter of common sense. The type of data and the reasons for requiring an average to represent it must be considered. Guidance is given under the section headings which follow. Generally it amounts to weighing the advantages and disadvantages with respect to ease of calculation, ease of understanding and avoidance of ambiguity or of any tendency to mislead, and suitability for further statistical work. For practical business use the arithmetic mean is usually favoured, while the sociologist makes great use of the median.

The equipment available for statistical calculations is a factor to be considered. Some calculations are especially onerous if done manually. A computer with an appropriate program may enable us to select the most suitable measure without reference to the cumbersome nature of the calculations and the greater probability of mistakes arising if a person had to do them.

## 3.2.1 ARITHMETIC MEAN

This is what the man in the street usually understands when he refers to an *average*. It is calculated by dividing the total of the data by the number of items involved, or the total frequency.

The total of the bonus data used in chapter 2 is £807·65, so the average bonus is $\frac{£807{\cdot}65}{150} = £5{\cdot}384$.

This calculation can be expressed as the formula $\mu = \frac{\Sigma x}{N}$. Here x denotes the variable being considered (the bonuses in the example) and N is the total number of items. The Greek capital sigma ($\Sigma$) is used in statistics to denote the sum of what follows. The right hand side of the equation implies that all the values of x are added and the total is then divided by the number of items. The Greek mu ($\mu$) has been used for the arithmetic average which results. If we wish to show that $\mu$ relates to x values we may use the form $\mu_x$; the mean of another set of values of, say, y would be denoted by $\mu_y$.

Other symbols may be used for the mean. An obvious letter is *m*, while we sometimes place a line over the x, giving $\bar{x}$ (read as bar-x or x-bar). Strictly the Greek symbol refers to the mean of the whole of the data, while the normal alphabet implies that we are using a sample from this. This is an important distinction at a later stage.

The fact that every person has some acquaintance with this form of average going back to school days is greatly in its favour. From a mathematical point of view it is the most useful form of average. It takes account of all the data, which is a strong point in its favour if calculations are required with the average as starting point.

Knowing the arithmetic average and the total frequency we can at once, by multiplication, calculate the total of the items. The effect of new items on the average can readily be estimated by adding them to this total and dividing the result by the new number of items.

Generally the calculation of the arithmetic mean is simple to perform and to understand, though it can be laborious with a large mass of data. No special arrangement of the data is needed. The arithmetic mean can be calculated directly from raw data, unlike the median and the mode. On the other hand, if the data are already arrayed or grouped the latter averages may be more readily found.

The arithmetic mean can sometimes give rather odd results. If we calculate the average number of persons in a family we may arrive at some figure, such as 3·24, which would be difficult to visualise. The arithmetic average is a mathematical concept and does not necessarily produce a representative item from the data.

The arithmetic average can give misleading impressions if the data really consists of two or more separate sets. If, for instance, we measured the heights of all people in a block of flats and found the arithmetic average we should be mixing adults and children and men and women and it would be difficult to interpret the average. This is not so much a criticism of the average as of the wrong use of an average for data that is not homogeneous.

In business we often have situations where a few items are of great significance and the bulk of items may be, individually, of little importance. The values of shareholdings of members of limited companies often follow this pattern. Again certain stock items may contribute much more to profit than others. In cases such as this the arithmetic average should be used with great care, if at all.

### 3.2.2 MEAN USING GROUPED DATA

Often we need to estimate the mean from the condensed form of the data available in a frequency distribution. This does not necessarily give precisely the same result as we should obtain if all the items were available but it may be the best we can do.

Table 3.1A shows the simple approach to this problem. What we require is an estimate of the total of the data. This can then be divided by the number of items to give the required arithmetic mean. If we assume that the items

Table 3.1. *Calculation of the mean from grouped data*

A. *From basic data*

| Bonus x | Frequency f | f.x |
|---|---|---|
| 3·5 | 6 | 21·0 |
| 4·5 | 35 | 157·5 |
| 5·5 | 75 | 412·5 |
| 6·5 | 33 | 214·5 |
| 7·5 | 1 | 7·5 |
| | 150 | 813·0 |

B. *From assumed mean of 5·5*

| Bonus x | Frequency f | Deviation of x from 5·5 *d* | fd |
|---|---|---|---|
| 3·5 | 6 | −2 | −12 |
| 4·5 | 35 | −1 | −35 |
| 5·5 | 75 | 0 | 0 |
| 6·5 | 33 | 1 | 33 |
| 7·5 | 1 | 2 | 2 |
| | 150 | | −47 |
| | | | +35 |
| | | | −12 |

C. *From assumed mean of 4·5*

| Bonus x | Frequency f | Deviation of x from 4·5 *d* | fd |
|---|---|---|---|
| 3·5 | 6 | −1 | −6 |
| 4·5 | 35 | 0 | 0 |
| 5·5 | 75 | 1 | 75 |
| 6·5 | 33 | 2 | 66 |
| 7·5 | 1 | 3 | 3 |
| | 150 | | −6 |
| | | | +144 |
| | | | +138 |

in each group have an average value half way through the group we can multiply this average by the respective frequency and total the products. Here the groups are 3—4, 4—5 etc. and the centres are respectively 3·5, 4·5 etc. The estimate of the total is £813·0 so that the estimate of the mean is

$\frac{£813·0}{150}$ = £5·42, which is slightly different from the true figure of £5·38.

When performing calculations of this type regard must be paid to where the centre of the group really lies.

With the bonus data *3—4* implies any item that is 3 and has not reached 4, i.e. from 3·00 to 3·99 and the centre is the average of these two values, i.e. 3·5.

Common sense tells us that the output data in table 3.2 should be treated in the same way, unless there is a good reason to suppose that only an exact number of tons can be produced.

If the groupings referred to individual numbers (integers), as might happen if the data were the number of items withdrawn from stores, then 60—69 means precisely what it says, and the centre is $64\frac{1}{2}$ $\left(\text{i.e. } \frac{60+69}{2}\right)$.

A case of some importance is that of ages. The usual way of calculating these means that a person is 69 until his 70th birthday, so that the average age of group 60—69 would be 65 years.

Care must also be taken if groups of a width differing from the normal for the distribution are used, since their centre values will not form part of the regular progression.

Sometimes the beginning and end of a distribution are left open, e.g. the first item might be stated as being up to a certain value and the last item as being over a stated value. In this case judgment has to be used to provide a reasonable figure for the working centres. Usually it is satisfactory to assume these groups are of normal width. Really, if this assumption is badly at fault, some indication should have been given in the table that there are a significant number of items at a considerable distance from the main distribution.

When grouped data are used the formula for the mean becomes $\mu = \frac{\Sigma f.x}{\Sigma f}$, where f denotes the frequencies of each of the groups. We multiply each value of x by its frequency and sum the products. The result is divided by the sum of the frequencies.

### 3.2.3 WORKING FROM AN ASSUMED MEAN

Calculation of the arithmetic mean from a frequency distribution can often be simplified by working from an assumed mean, which is then adjusted to give the required value.

This is illustrated in table 3.1B. The first two columns are from the original distribution. The third column (headed *d* for *deviation*) shows by how much each item differs from an assumed mean of 5·5. Thus, the first item is 3·5 less 5·5 which is the value *—2* shown in the third column.

Column *d* items are then multiplied by the relevant frequencies to give the fourth column (headed fd). This column is totalled algebraically (i.e. taking account of the plus and minus signs), giving a value of —12. The implication of this value is that it represents the total effect on the sum of the deviations of the items due to 5·5 not being the exact mean. If it had been the exact mean the positive and negative products in the fd column would have cancelled out giving a total of zero. Division by 150, the number of items, gives the net effect. This amount must be added to (if positive) or deducted from (if negative) the assumed mean.

The result is $5.5 - \frac{12}{150} = 5{\cdot}5 - 0{\cdot}08 = 5{\cdot}42$, which is precisely the same result as we obtained by the longer method.

It does not affect the final result if a different assumed mean is taken but the amount of calculation may vary. Table 3.1C shows the result of using 4·5 as the basis of calculation. In this case the mean is obtained from

$$4{\cdot}5+\frac{138}{150}=5.42.$$

The original calculation, based on the actual values of the data, is really equivalent to working from an assumed mean of zero.

The bonus data are arranged in groups the width of which is unity. There is £1 difference between the centres of each group. In many cases this is not the case but the above method may still be used. The deviations are expressed in *working units* which happen to suit the case in hand and an adjustment is made, at the end, to the total of the deviations. This is sometimes called *coding* the data.

Table 3.2. *Calculation of mean from grouped data where group width (or interval) is not unity*

*Weekly output of factory over 50 weeks (in tons)*

| Output x | Frequency f | Deviation from 95 in 10's d | fd |
|---|---|---|---|
| 60— 69 | 2 | −3 | −6 |
| 70— 79 | 6 | −2 | −12 |
| 80— 89 | 10 | −1 | −10 |
| 90— 99 | 23 | 0 | 0 |
| 100—109 | 8 | 1 | 8 |
| 110—119 | 1 | 2 | 2 |
| | 50 | | −28 |
| | | | +10 |
| | | | −18 |

Table 3.2 illustrates this point. The output figures are in groups of width 10 units. The deviations could be expressed directly as −30, −20, etc. but it is easier to divide by the common factor of 10, resulting in deviations of −3, −2, etc. The total of the fd column, −18, has to be multiplied by 10, making −180, to give the true result.

The average output is thus $95-\frac{18\times 10}{50}=91.4$ tons.

### 3.3.1 THE MEDIAN

This presents an entirely different approach to the problem of selecting a single figure to represent the data. The median is simply the *middle item of an array*. Thus, if the data are in order, the median may be picked out directly.

The median of the numbers 26, 34, 77, 85, 99, is 77, the 3rd out of the 5 items. The rule is to take the $\frac{N+1}{2}$ item, when each item is available and the series is not grouped or continuous. Thus, if there are 167 items in a series, the median is the 84th. If the series has an even number of items then the rule is to take the mean of the two middle items. With the bonus data the median is the $75\frac{1}{2}$th item, i.e. the average of the 75th and the 76th. In this case both happen to be the same, so that the median is immediately found to be £5·5

The advantages of such an easily derived average are obvious from the point of view of calculation. It is also easy to understand and to explain to people who are not mathematically inclined. It has the additional merit of being very often an actual item, which has more meaning than some of the more abstruse averages. It may be unnecessary actually to measure all the items in the series so long as they can be put in order. If we wished to find the median height of workers in an office we could arrange for them to stand in order of height, pick out the middle person, and measure him.

The fact that the median depends on the central placing of an item and does not take account of the absolute size of the other items is sometimes an advantage. Odd items departing very much from the general run of the data do not affect the actual size of the median, as they would if the arithmetic average were involved. The median would be more realistically located among the bulk of the items.

This merit from the point of view of understanding is sometimes a disadvantage mathematically. For further work it may be desirable to have an average that incorporates all the data.

### 3.3.2 GRAPHICAL INTERPOLATION OF THE MEDIAN

With data that are grouped the median is slightly more difficult to find. Normally it occurs part of the way through a group and we have to estimate the appropriate point. This process of selecting a value between existing values is called *interpolation* and is a very important statistical process, quite apart from its use in connection with the median.

The cumulative frequency table (2.7.1) and the ogive (2.7.2) based on it can be used for interpolation.

The output data have been treated in this way in table 3.3 and in diagram 3.1, based on the ascending figures.

Table 3.3. *Cumulative frequencies of output data*

| Output | Frequency | Cumulative frequency | |
|---|---|---|---|
| | | ascending | descending |
| 60— 69 | 2 | 2 | 50 |
| 70— 79 | 6 | 8 | 48 |
| 80— 89 | 10 | 18 | 42 |
| 90— 99 | 23 | 41 | 32 |
| 100—109 | 8 | 49 | 9 |
| 110—119 | 1 | 50 | 1 |

A line is drawn across at the median frequency and the value of x at which this line cuts the curve is read off. Our graph shows the value of the median to be about 93 tons.

It should be noted that, for data that are grouped, the median is the $\frac{N}{2}$ item, instead of $\frac{N+1}{2}$. The reason for this is that a different value would be obtained using an ascending ogive from using a descending one. Thus, with 50 items, $\frac{50+1}{2}=25\frac{1}{2}$. If we deduct this from 50 we get $24\frac{1}{2}$. With discrete, or individual values, as previously considered, counting back from 50 as the first number gives $25\frac{1}{2}$, the same as if we started from the lower end.

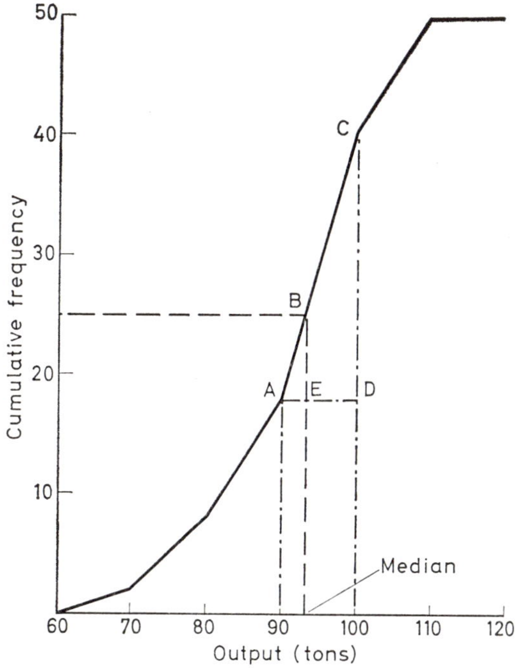


Diagram 3.1. *Median* by interpolation from the ogive

### 3.3.3 CALCULATING THE MEDIAN

This method is equivalent to interpolating from the ogive but the result is achieved by calculation and the graph is unnecessary.

The median is the 25th item. In table 3.3 we look down the cumulative frequency column (ascending) to find the first cumulative frequency that is greater than this value. Clearly the median must occur in the group in which this cumulative figure occurs. In the case of the output data the first cumulative frequency above 25 is 41, and the group is 90—99.

The value we require is a fraction of the way through this group. If we assume the data are equally spread through the group we can estimate this fraction by taking the number we have to go into the group to find the median and dividing this by the group frequency. In the case considered the distance into the group is $25-18=7$, since 18 is the number of items occurring before this group. The group frequency is 23, so that the fraction through the group is $\frac{7}{23}=0{\cdot}304$.

We multiply this by the group width of 10 to arrive at a value of 3·04 which must be added to the lower limit of the group (90) to give the required median as 93·04.

This amounts to the following calculation:

$$\text{median}=90+\frac{10\,(25-18)}{23}=93{\cdot}04 \text{ tons.}$$

Just to check, using the descending data, we would get, median$=100-\frac{10\,(25-9)}{23}=93{\cdot}04$. Notice that, in this case, we work from the tops of the groups and have to deduct the amount involved, remembering that the top of the relevant group is really 100 and not 99.

Diagram 3.1 explains this process in terms of the ogive. We can consider the similar triangles ABE and ACD. $\frac{AE}{AD}=\frac{BE}{CD}$ so that AE, the amount required to add to 90, is $AD.\frac{BE}{CD}=\frac{10\,(25-18)}{23}$, agreeing with the result above.

## 3.3.4 PARTITION VALUES

The median, occurring half way through the data, may be regarded as a very important example of what are referred to as *partition values.* These are values that divide the data into some predetermined fractions.

The best-known partition values are:

a. The quartiles, dividing the data into quarters. The first and third of these are usually denoted by $Q_1$ and $Q_3$ or sometimes by $Q_L$ and $Q_U$, denoting lower and upper respectively. The second quartile corresponds to the median.
b. The quintiles, dividing the data into fifths.
c. The deciles, dividing the data into tenths.
d. The percentiles, dividing the data into hundredths, of which selected examples are chosen.

The methods of estimation of the partition values are precisely the same as for the median.

The upper quartile of the output data is the $\frac{3}{4}\times 50=37{\cdot}5$th item. This occurs again in the 90—99 group and is given by $90+\frac{10\,(37{\cdot}5-18)}{23}=98{\cdot}5$ tons.

## 3.4.1 MODE

The mode may be defined simply as *the item that occurs most frequently.*

For the bonus data study of table 2.1 shows that this item is 5·75, which appears fourteen times.

To be of use the mode must convey in itself a picture of the mass of data from which it is drawn. With an array of individual items it would be possible for a relatively large or small item to occur most often and to give a misleading picture.

It is best to think in terms of a frequency distribution of grouped data. If this is displayed in histogram form and shows an appreciable hump this may indicate where the mode lies. It is often more convenient to deal with the *modal group* (the group having the largest frequency) than with a precise mode.

From table 2.4 the modal group is seen to be the 5·5—6 group. From table 2·5 it is the 5—6 group. From table 3·2 it is the 90—99 group.

When a mode is clear in the data it can be a useful form of average. It is easy to understand and easy to obtain, especially with grouped data. It has the supreme merit over other averages that it can be obtained when some characteristic cannot itself be measured, such as the colours of cars that people prefer.

Again, when something that is measurable can occur only in predetermined values, such as shoe sizes, the manufacturer is more concerned with which of these are most popular than with an arithmetic average which may come out at some size which does not exist.

The mode is a useful, popular measure in such cases, but it is not as valuable as the arithmetic mean as a basis for further statistical work with the data. Like the median it has the practical merit of being unaffected by extreme items.

Many distributions met with in practice will be found to have a clearly defined mode. A very important pattern is that of the Gaussian curve dealt with in chapter 8. There is no reason to assume, though, that every distribution will conform to this pattern with a pronounced, centrally-placed mode. With continuous data we may regard the mode as the point at which the maximum value of the probability density function occurs (see section 2.5.3).

### 3.4.2 BIMODAL DISTRIBUTIONS

Sometimes a distribution appears to have more than one mode. If two occur it is referred to as *bimodal.*

This odd shape may lead us to suspect that two or more sets of data have been merged. If two machines are set to produce the same dimension of output there will probably be some difference between the sets produced on each machine. If each has a different mode the effect of combining their two outputs may be to display these two modes in a bimodal distribution. A similar result occurs if the setting of one machine slips significantly. This can cause difficulties in the application of theories, such as those relevant to statistical quality control, which assumes that all data are from one source.

Table 3.4. *Dimensions of the same product from two machines (A and B)*

| Diameter (inches) | % of output | | Combined effect | |
|---|---|---|---|---|
| | A | B | Total | % |
| 0·995—0·997 | 4 | — | 4 | 2 |
| 0·997—0·999 | 15 | — | 15 | 7·5 |
| 0·999—1·001 | 65 | 4 | 69 | 34·5 |
| 1·001—1·003 | 13 | 16 | 29 | 14·5 |
| 1·003—1·005 | 2 | 70 | 72 | 36 |
| 1·005—1·007 | 1 | 8 | 9 | 4·5 |
| 1·007—1·009 | — | 2 | 2 | 1 |
| | 100 | 100 | 200 | 100 |

An example is shown in table 3.4 which illustrates the effect of combining the outputs of two machines, shown as percentages. For the percentages to be added directly like this the quantity produced by each machine would have to be the same.

### 3.4.3 CALCULATING THE MODE

We can interpolate within the modal group to obtain an estimate of the mode. This may be at the expense of one of the chief features of the mode, that it is an actual item taken from the data. For statistical analysis, however, we are sometimes concerned with finding whether the mode of data has shifted, perhaps with the passage of time.

A suitable formula is:

$$\text{Mode} = L + \frac{I \times (f_M - f_L)}{(2f_M - f_L - f_U)}$$

where L is the lower limit of the modal group,
$f_M$ is the frequency of the modal group,
$f_L$ is the frequency of the group below the modal group,
$f_U$ is the frequency of the group above the modal group, and
I is the group interval or width.

Application of this formula to the data of table 3.3 results in:

$$\text{Mode} = 90 + 10 \times \frac{(23-10)}{(46-10-8)} = 94{\cdot}6 \text{ tons.}$$

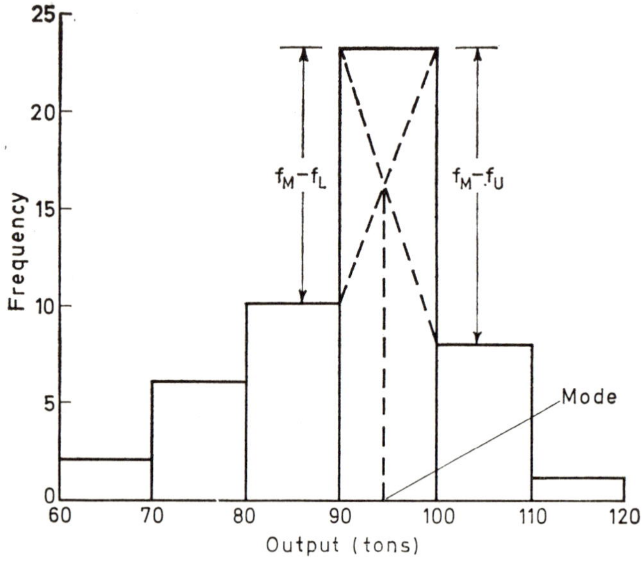


Diagram 3.2. *Mode* with grouped data

Diagram 3.2 shows a simple graphical construction for arriving at the mode. The reader should check that this is the equivalent of what is achieved by the formula. Consider similar triangles and the fact that $2f_M - f_L - f_U = (f_M - f_L) + (f_M - f_U)$.

Another approach where the mode is not obvious is to group the data in different ways. This regrouping may create a new modal group. Formation of broader groupings may also achieve this.

## 3.5 THE GEOMETRIC MEAN

The simple mathematical interpretation of the geometric mean is that, if we have two numbers such as 3 and 12, their geometric mean is the number x, such that $\frac{3}{x} = \frac{x}{12}$. This implies that $x = \sqrt{36} = 6$.

Extension of this principle for use as an average implies that we should *multiply the numbers together and take the root equivalent to the total frequency.* For N numbers we take the Nth root of the product of the N numbers.

Table 3.5. *Calculation of the geometric mean*

| Number | Log. |
|---|---|
| 6·5 | 0·8129 |
| 2·0 | 0·3010 |
| 4·4 | 0·6435 |
| 10·2 | 0·0086 |
| 0·8 | $\bar{1}$·9031 |
| | 5)1·6691 |
| | 0·3338 |

Antilog of 0·3338=2·157, which is the G.M.

The practical way of doing this is to add together the logarithms of the numbers and divide the total of the logarithms by the number of items. The antilog of this value then gives the required geometric mean.

This is shown in table 3.5. It will be seen that the calculation amounts to finding the arithmetic average of the logarithms of the numbers and looking up the antilog of this.

Table 3.6. *Calculation of geometric mean when frequencies are involved*

| Item | Cost (£) | Frequency (f) | Log. Cost | f× log. |
|---|---|---|---|---|
| A | 6·5 | 3 | 0·8129 | 2·4387 |
| B | 2·0 | 6 | 0·3010 | 1·8060 |
| C | 4·4 | 10 | 0·6435 | 6·4350 |
| D | 10·2 | 6 | 1·0086 | 6·0516 |
| E | 0·8 | 5 | $\bar{1}$·9031 | $\bar{1}$·5155 |
| | | 30 | | 30)16·2468 |
| | | | | 0·5416 |

Antilog. of 0·5416=3·48, which is the G.M.

When each number does not stand alone but has a frequency to be taken into account we must multiply each logarithm by this frequency. We then add the products and divide the total by the sum of the frequencies. The geometric mean is the antilog of this sum.

A calculation of this type is shown in table 3.6.

The demerits of such an average from the points of view of difficulty of calculation and comprehension are obvious. It has certain advantages in the case of index numbers (see chapter 5).

## 3.6 THE HARMONIC MEAN

This average is best explained by means of a practical example.

If we drive equal distances at 20 m.p.h. and at 30 m.p.h. the average speed is not the average of these two speeds. At least it is not the arithmetic average, which is 25 m.p.h. A drive of 60 miles at 20 m.p.h. takes 3 hours. A similar distance at 30 m.p.h. takes 2 hours. Thus we have travelled 120 miles in 5 hours, giving an average speed of 24 m.p.h.

This can be called the harmonic mean of 20 and 30 and can be derived as follows:

$$\text{Harmonic mean} = \frac{2}{\left(\frac{1}{20}+\frac{1}{30}\right)} = 24$$

In general terms the harmonic mean $= \dfrac{\Sigma f,}{\Sigma\left(\frac{f}{x}\right)}$ where the various items, denoted by x, occur with frequencies denoted by f. We *divide the total number of items by the sum of the reciprocals of the items*, each multiplied by its appropriate frequency.

A practical example of this is given in table 3.7, which relates to a journey done at different speeds for varying distances. In such a case the miles travelled at each speed may be regarded as the frequencies for these speeds.

Table 3.7. *Vehicle speeds and distances covered*

| Speed (m.p.h.) | Distance (miles) | Time (hrs) |
|---|---|---|
| 10 | 10 | 1 |
| 20 | 10 | 0·5 |
| 30 | 90 | 3 |
| 40 | 80 | 2 |
| 50 | 75 | 1·5 |
| 60 | 30 | 0·5 |
| 70 | 35 | 0·5 |
| | 330 | 9·0 |

$$\text{Average speed} = \frac{330}{9} = 36{\cdot}7 \text{ m.p.h.}$$

$$\text{H.M. of speeds} = \frac{330}{\frac{10}{10}+\frac{10}{20}+\frac{90}{30}+\frac{80}{40}+\frac{75}{50}+\frac{30}{60}+\frac{35}{70}} \text{ m.p.h.}$$

$$= 36{\cdot}7 \text{ m.p.h.}$$

Another way of expressing the harmonic mean is that it is the reciprocal of the arithmetic average of the reciprocals of the items.

### 3.7.1 WORKED EXAMPLE

*Question*

From the age distribution data of table 2.8 calculate:

*a.* The arithmetic mean age.

*b.* The median.

*c.* The lower quartile.

*Answer*

*a.* [This is an obvious case for working from an assumed mean, which has been taken as 35. Table 3.8 shows the accumulation of the necessary figures.]

Table 3.8. *Age distribution of U.K. (1961)*

| Age | Number (f) | Age mid-pt | Deviation (d) from 35 in 10's | f.d | Cum.f |
|---|---|---|---|---|---|
| 0— 9 | 8,028 | 5 | −3 | −24,084 | 8,028 |
| 10—19 | 8,003 | 15 | −2 | −16,006 | 16,031 |
| 20—29 | 6,563 | 25 | −1 | −6,563 | 22,594 |
| 30—39 | 7,083 | 35 | 0 | 0 | 29,677 |
| 40—49 | 7,100 | 45 | 1 | 7,100 | 36,777 |
| 50—59 | 6,958 | 55 | 2 | 13,916 | 43,735 |
| 60—69 | 5,019 | 65 | 3 | 15,057 | 48,754 |
| 70—79 | 2,937 | 75 | 4 | 11,748 | 51,691 |
| 80—90 | 1,017 | 85 | 5 | 5,085 | 52,708 |
| | 52,708 | | | −46,653 | |
| | | | | +52,906 | |
| | | | | +6,253 | |

$$\text{Arithmetic average age} = 35 + 10 \times \frac{6{,}253}{52{,}708}$$
$$= 35 + 1{\cdot}19$$
$$= \mathit{36{\cdot}2\ years}$$

[It is pointless pretending to too great accuracy in cases such as this. We are really only justified in saying that the mean is approximately 36 years.]

*b.* The median is the $\frac{52{,}708}{2} = 26{,}354$th item.

It occurs in the 30—39 year group and is $30 + 10 \times \frac{(26{,}354 - 22{,}594)}{7{,}083}$

$= \mathit{35{\cdot}3}$ *years.*

*c.* The lower quartile is the $\frac{52{,}708}{4} = 13{,}177$th item.

It occurs in the 10—19 year group and is $10 + 10 \times \frac{(13{,}177 - 8{,}028)}{8{,}003}$

$= \mathit{16{\cdot}4}$ *years.*

## 3.7.2 WORKED EXAMPLE

*Question*

From the Trade Union data shown in table 3.9 calculate:

*a.* The median membership.

*b.* The arithmetic average membership.

*c.* The average membership of the 18 largest unions.

Also comment on the difference between the median and the arithmetic average.

Table 3.9. *Trade Unions in U.K. (1964)*

| Membership | No. of Unions | Total Membership (thousands) |
|---|---|---|
| under 100 | 109 | 5 |
| 100— 499 | 143 | 36 |
| 500— 999 | 55 | 38 |
| 1,000— 2,499 | 100 | 161 |
| 2,500— 4,999 | 60 | 207 |
| 5,000— 9,999 | 28 | 189 |
| 10,000—14,999 | 22 | 271 |
| 15,000—24,999 | 20 | 388 |
| 25,000—49,999 | 19 | 706 |
| 50,000—99,999 | 17 | 1,196 |
| 100,000 and over | 18 | 6,869 |
| | 591 | 10,066 |

(*Dept. of Empl. and Prod.*)

*Answer*

*a.* [Table 3.10 shows the cumulative figures needed to derive the median. Strictly it would be necessary only to go as far as the group that gives a cumulative figure greater than the median cumulative frequency.]

Table 3.10. *Trade unions in U.K. (1964)*

| Membership | No. of Unions | Cumulative No. |
|---|---|---|
| under 100 | 109 | 109 |
| 100— 499 | 143 | 252 |
| 500— 999 | 55 | 307 |
| 1,000— 2,499 | 100 | 407 |
| 2,500— 4,999 | 60 | 467 |
| 5,000— 9,999 | 28 | 495 |
| 10,000—14,999 | 22 | 517 |
| 15,000—24,999 | 20 | 537 |
| 25,000—49,999 | 19 | 556 |
| 50,000—99,999 | 17 | 573 |
| 100,000 and over | 18 | 591 |

The median is the $\frac{591}{2}$=295·5th item.

It occurs in group 500—999, and is $500+500\times\frac{(295\cdot5-252)}{55}$

$=500+395$

$=$*895 members.*

*b.* [Here there is no need to multiply the centres of the groups by their frequencies since we are told the accurate total membership. Division of this by the number of unions immediately gives the required average.]

Arithmetic average $=\frac{10{,}066{,}000}{591}$

$=$ *17,000 members.*

*c.* Average membership of 18 largest unions $=\frac{6{,}869{,}000}{18}$

$=$ *382,000 members.*

The difference between the averages is accounted for by the uneven distribution of the data. The 18 largest unions account for over 60% of total union membership, while the 307 smallest (109+143+55), amounting to over 50% of the number of unions, have a membership of only 79,000 (5+36+38), which is less than 1% of total membership. The arithmetic average depends on the total membership, the median on the ranking.

## 3.8 FURTHER QUESTIONS

1. The arithmetic average of 150 values of x is 3·7. A further 5 values (4·2, 6·8, 5·0, 7·1, 6·2) become available. Calculate the new arithmetic average, including these values.

2. Give five examples each of practical situations where *a.* the mode, and *b.* the median would be the most suitable average to use, briefly stating reasons.

3. For the data of table 3.11 calculate for all establishments:
   *a.* The median size.
   *b.* The arithmetic average size.
   Comment on your results.

Table 3.11. *Size of manufacturing establishments in G.B. June, 1961 (over 10 employees)*

| No. of Employees | No. of Establishments All | Vehicles | Textiles | (Total of Employees (000's) All | Vehicles | Textiles |
|---|---|---|---|---|---|---|
| 11— 24 | 12,571 | 388 | 875 | 222 | 7 | 16 |
| 25— 99 | 27,478 | 810 | 2,568 | 1,420 | 41 | 141 |
| 100— 499 | 12,213 | 437 | 1,878 | 2,552 | 92 | 403 |
| 500— 999 | 1,693 | 120 | 184 | 1,163 | 84 | 124 |
| 1,000—1,999 | 777 | 82 | 38 | 1,078 | 115 | 56 |
| 2,000 or more | 429 | 94 | 16 | 1,743 | 515 | 49 |
| | 55,161 | 1,931 | 5,559 | 8,178 | 854 | 789 |

(*Dept. of Empl. and Prod.*)

4. Calculate the lower and upper quartile sizes of establishments in the vehicle industry in 1961, from data in table 3.11.

5. Calculate the arithmetic average size of textile establishments from table 3.11, assuming the last column, showing the totals of employees, is not available. Check your estimate, using this column.

6. Comment on the differences displayed in table 3.11 between the three sets of figures.

7. Comment on this statement: 'It is misleading to say that the average wage of workers in this industry is £30 a week. At least half of the workers receive less than that.'

8. Calculate the average diameter of the outputs of the two machines A and B shown in table 3.4.
   What will be the average of the combined data?

9. Calculate the geometric mean of the first ten even numbers.

10. A person regularly invests £20 a month in a unit trust. For the first five months the prices of the units bought were £0·5, £0·6, £0·65, £0·7, £0·6. Use a statistical average to calculate the average purchase price of the units bought and check your result by an ordinary arithmetic approach.

# 4

# Approximation and Errors

## 4.1 SOURCES OF ERROR

Errors may occur in statistics for a number of reasons.

Some are deliberately introduced, as, for example, when data is rounded off or approximated, perhaps to make it easier to grasp or to simplify calculations.

Most measurements may be regarded as approximations, the accuracy being related to the precision of the measuring instrument used and the skill of its user.

Frequencies in statistical surveys, when the number included is large, are suspect. The errors may arise through miscounts, though the work should be organised to show up any serious errors arising from this cause, for example by check totals from different sources. At elections, when the results are close, recounts usually produce different figures, in spite of the precautions taken. Another source of such errors is the misinterpretation of the forms to be completed and the questions asked.

Errors may also be introduced by deficiencies in the computational methods or equipment used. We may speak of *slide rule accuracy* or may use a formula that ignores minor features of the data.

Most statistical work uses samples and *sampling errors* arise from the fact that different samples tend to give different results (see chapter 9). They are often called *statistical errors.*

Increases in accuracy have to be paid for in time or money, or both, and we normally compromise at a degree of accuracy that is not too costly when compared with what is achieved.

## 4.2 ROUNDING AND ITS EFFECTS

*Rounding* implies that the figures are simplified and may be done in a number of ways.

The easiest approach is to ignore completely the less important parts of the number. If we take the items in table 1.1, then 5·30 becomes 5, if we are *rounding down* to units, or 5·3, if we are rounding down to the first decimal; 3·05 becomes 3 or 3·0 etc. This method always results in something being discarded, unless the original figure is the same as the rounded one.

As more rounded numbers are added the *absolute error* in the total becomes larger. We can estimate the likely effect of this *bias* in approximation on the total by considering that, for each column of figures discarded, numbers from 0 to 9 have an equal chance of occurring. For each item the average loss will be $4\frac{1}{2}$ of the units of that column, in the long run. If we check this, using the first column of 25 items and discarding the hundredths column, we find the total is 121·8. The expected error is $25 \times 0{\cdot}045 = 1{\cdot}125$. Adding this to the total gives $121{\cdot}8 + 1{\cdot}125 = 122{\cdot}925$.

If the original figures are added the total is found to be 122·45. A check through the items shows that, in this case, the assumption that any digit has an equal chance of being discarded was wrong. Only 0 and 5 seem to occur in the relevant column anywhere in the table, so a better approximation to the likely error would have been the average of 0 and 5, that is $2\frac{1}{2}$, multiplied by 25, giving an error in the total of $2{\cdot}5 \times 25 = 62{\cdot}5$ hundredths, or 0·625. An estimate of the correct total is then $121{\cdot}8 + 0{\cdot}625 = 122{\cdot}425$, which is very close to the actual total.

The opposite to rounding down is *rounding up*. Anything above zero following the figure to which we are rounding causes it to be increased by one. The effect on totals resembles that of rounding down, except that it will increase the totals.

The most usual method of rounding is to do this to the *nearest unit* being considered. Figures from 0 to 4 are discarded, those from 6 to 9 cause the figure being approximated to be increased by 1. If 5 occurs this is commonly rounded upwards; an alternative is to round up or down to an even figure, (e.g. 0·75 becomes 0·8 and 0·85 also becomes 0·8) so that this centre value tends to cancel out in the long run.

Rounding the numbers in the first column of table 2.1 to the nearest tenth results in 5·3, 3·1 (if we round up) or 3·0 (with even rounding), 4·8 (whether we round up the 5 or round to an even figure), 3·6, 6·1, 3·8, 4·1, 6·3 or 6·2 etc.

The effect of such rounding on long-run totals is to cancel out, if the even figure rounding of 5 is used, since 1, 2, 3, and 4 are discarded, 5 evens out, and 6, 7, 8, and 9 cause increases of 4, 3, 2, and 1 to be made. If 5 is rounded upwards there will be a tendency for an increasing error in the total. With the given data, since only 0 or 5 can occur the error in 25 items would tend to $25 \times 2\frac{1}{2} = 62{\cdot}5$ or 0·625, if we are rounding to tenths.

The *relative error* may be of more interest than the absolute error. It is the absolute error represented as a percentage of the figure to which it relates.

Rounding up and down suffer from *bias* or a tendency to sway the data in a particular direction. This is to be avoided, if possible.

Sometimes we may be interested in estimating the *maximum* error that could occur, rather than the *probable* error that we have been dealing with above. In the case of rounding up the maximum extent of this would be to add on 9 units for each item. For the data this would give a maximum error in total of $25 \times 0{\cdot}09 = 2{\cdot}25$.

In computer calculations it is important to distinguish between *integers* or whole numbers, and *real numbers*, which can have a decimal part. The result of dividing 3 by 2 may appear as 1, since the decimal will be ignored, if the numbers, and the quotient, have been specified as integers.

When figures have been rounded this should be indicated. An opportunity often arises for reducing the space occupied by data by selecting units that will eliminate zeros. Thus, if nothing less than a million occurs, 6 zeros may be omitted and the data shown as *in millions*, or *000,000 omitted*, or $\times 10^6$. This method of using powers of 10 is generally accepted mathematically.

When decimals occur, such as in 45·0168, it is sometimes useful to express them as negative powers of 10, e.g. $450168 \times 10^{-4}$. This is especially useful for numbers such as 0·000000036, which becomes $36 \times 10^{-9}$, and is much easier to handle. This, of itself, does not involve approximation but is often coupled with it.

It is sometimes useful to present all numbers in a standard form. Alternatives are to have one digit to the left of the point or to standardise each number as a decimal; in either case the power of 10 needed to give the original number is stated.

On the first system $3{\cdot}87 \times 10^5$ would be 387,000 and $3{\cdot}87 \times 10^{-2}$ would be 0·0387. On the second system we should have $0{\cdot}387 \times 10^6$ and $0{\cdot}387 \times 10^{-1}$ for these numbers.

## 4.3.1 ADDITION

If two numbers, each of which may vary between certain limits, are added, the effect is that the total can vary between the sums of the respective lower and upper limits.

Thus, if the labour costs of a factory for a period are estimated to be £10,000±1,000 (i.e. can vary from 9,000 to 11,000) and the materials costs for the same period are estimated to be £6,000±750 (i.e. from 5,250 to 6,750) then the following applies:

$$\begin{aligned} \text{estimated total} &= 10{,}000+6{,}000=16{,}000 \\ \text{least total} &= 9{,}000+5{,}250=14{,}250 \\ \text{greatest total} &= 11{,}000+6{,}750=17{,}750 \end{aligned}$$

These statements are equivalent to stating that the total=16,000±1,750 where 1,750=1,000+750.

Sometimes the possible error may differ, depending on whether it is above or below the figure estimated. The labout data might be £10,000 $^{+2,000}_{-1,000}$ and the materials £6,000$^{+800}_{-500}$ the effect of which would be:

$$\begin{aligned} \text{estimated total} &= 10{,}000+6{,}000=16{,}000 \\ \text{least total} &= 9{,}000+5{,}500=14{,}500 \\ \text{greatest total} &= 12{,}000=6{,}800=18{,}800 \end{aligned}$$

which is equivalent to stating that the total=16,000$^{+2,800}_{-1,500}$ where 2,800=2,000+800 and 1,500=1,000+500.

The procedure is thus to add the estimated figures to give the estimated answer. The limits are found simply by taking the sums of the lower and upper limits.

The total range of error can be found by adding together all the possible errors, i.e. 2,000+1,000+800+500=4,300=18,800−14,500.

It should be noted that if the errors are expressed as percentages the percentages cannot be added in this way. Thus, in the first example, the percentage error for the labour estimate is $10\frac{1}{2}\%$ and for the materials $12\frac{1}{2}\%$. The percentage error in the total is $\frac{1{,}750 \times 100}{16{,}000} = 10{\cdot}9\%$.

### 4.3.2 SUBTRACTION

The same considerations apply as in addition. If the income of a business is £10,000$\pm$1,000 and costs are £6,000$\pm$750 then

$$\text{estimated profit} = 10{,}000 - 6{,}000 = 4{,}000$$
$$\text{least profit} = 9{,}000 - 6{,}750 = 2{,}250$$
$$\text{greatest profit} = 11{,}000 - 5{,}250 = 5{,}750$$

or the profit estimate$=$4,000$\pm$1,750.

It will be noted that if the two items being considered are nearly equal the effect of the two errors in the estimates, being additive, may be relatively large compared with the estimated answer. In this case the error is $\frac{1{,}750}{4{,}000} \times 100 = 43\tfrac{3}{4}\%$. The significance of this to businessmen, whose success depends ultimately on this simple subtraction sum, should not need stressing.

### 4.3.3 MULTIPLICATION

If we multiply 10$\pm$3 by 20$\pm$2 we have the following results

$$\text{estimated product} = 10 \times 20 = 200$$
$$\text{least product} = 7 \times 18 = 126$$
$$\text{greatest product} = 13 \times 22 = 286$$

or the product is $200^{+86}_{-74}$ or $200^{+43\%}_{-37\%}$.

The calculation amounts to

$$(10\pm3)\times(20\pm2) = 200\pm60\pm20+6$$

treating this as the expansion of $(x\pm a)\,(y\pm b) = xy \pm ay \pm bx + ab$.

In the example the errors are very large compared with the items being multiplied. If the estimated errors are relatively small the ab term becomes insignificant, and can be ignored. The rule then is to add together (first item$\times$error in the second item)$+$(second item$\times$error in the first item).

In the example this would give $(10\times2)+(20\times3)=80$ and the answer would seem to be 200$\pm$80.

It will be seen that this is a percentage error of $\frac{80}{200}\times100=40\%$ and with the products of approximated figures we can estimate the percentage error in the product by adding together the percentage errors in the original estimates.

### 4.3.4 DIVISION

The effect of dividing 100$\pm$10 by 20$\pm$5 is

$$\text{estimated result} = \frac{100}{20} = 5$$

$$\text{least result} = \frac{90}{25} = 3{\cdot}6$$

$$\text{greatest result} = \frac{110}{15} = 7{\cdot}3$$

which is equivalent to $5^{+2\cdot3}_{-1\cdot4}$.

If we think in terms of $\frac{x \pm a}{y \pm b}$ the difference between the greatest and the smallest result is $\frac{x+a}{y-b} - \frac{x-a}{y+b} = \frac{2(ay+bx)}{y^2}$ if the very small terms are ignored.

This is twice the error possible if the items were multiplied, divided by the square of the divisor.

In the example above it is $2\,\frac{(100 \times 5 + 20 \times 10)}{400} = 3{\cdot}5$, compared with the actual result of $2{\cdot}3 + 1{\cdot}4 = 3{\cdot}7$.

In the example chosen the errors associated with the two items are relatively large. If the numbers had been $100 \pm 2$ and $20 \pm 1$ the results would have been, greatest 5·37 and least 4·67, which are fairly evenly spread on each side of 5. Use of the formula gives the total error as approximately 0·7, and we can divide this by 2 to give the upper and lower limits.

### 4.3.5 SQUARES

This is a special case of multiplication where both numbers are the same. The error possible is twice the number multiplied by the error in it, above or below the estimated squared value (i.e. $x \pm 2ax$).

## 4.4 WORKED EXAMPLE

*Question*

A company is marketing a new product, which it intends to sell at £2. The sales for the first year are estimated to be 100,000±10%. The costs are calculated to be made up of two parts, a fixed overhead portion estimated at £90,000±5%, and a direct cost for each unit of £0·8, which it is thought may increase by as much as 20%. The product will be manufactured against actual orders.

Show the range of profits that might arise.

*Answer*

| | | |
|---|---|---|
| Expected sales revenue = 100,000 × £2 | = | £200,000 |
| Expected costs of these sales | | |
| = £90,000 + 100,000 × £0·8 | = | £170,000 |
| Giving a profit of ............................... | | £30,000 |
| Least sales revenue = 90,000 × £2 | = | £180,000 |
| Greatest cost of 90,000 units (including the 5% and 20% increases) | | |
| = £94,500 + 90,000 × £0·96 | = | £180,900 |
| Giving a loss of ............................... | | £900 |
| Greatest sales revenue = 110,000 × £2 | = | £220,000 |
| Least cost of 110,000 units (deducting 5% from overheads) | | |
| = £85,500 + 110,000 × £0·8 | = | £173,500 |
| Giving a profit of ............................... | | £46,500 |

The profits may thus range from a loss of £900 to a maximum profit of £46,500, with an estimate of £30,000 profit.

## 4.5 PRACTICE QUESTIONS

1. Comment on the errors possible in the calculation of the arithmetic average membership of unions in worked example 3.7.2.

2. After taking stock a firm arranges for the items to be priced and then leaves the extension of the quantities × prices and the additions of these products to comptometer operators (see 19.2). State some of the errors which might arise and methods you would suggest to check the final accuracy of the stock value.

3. Discuss some of the main errors which may arise in connection with census data collection and suggest precautions to be taken.

4. Distinguish between sampling errors and other forms of error. [This question will be better answered after reading chapter 9.]

5. Explain the difference between absolute and relative error. Discuss whether it is advisable in business to rely exclusively on either of these.

6. A man is trying to estimate the amount of concrete needed for the concrete base of a new garage. He estimates that this will be about $16' \times 10'$, 'give or take a foot in each case' and that it will be about $6''$ thick, 'give or take an inch'. Calculate the absolute and relative error possible in the volume.

# 5

# Index Numbers

## 5.1 INTRODUCTION. SINGLE ITEMS

An *index number* attempts to show how some item or group of items differs from a standard or base. The most frequent use is in connection with time series, where the comparison is with a selected year, termed the *base year*, and usually takes the form of a *percentage*.

Typical index numbers are concerned with retail prices, wholesale prices, volumes of output, exports, productivity, wage rates, and business activity.

The simplest type of index number arises when there is only one item. If a firm makes only one product and the output of this for three years is 90, 120 and 150 tons, respectively, then we can simply select a base year and express the outputs for the other years as percentages of this.

If the first year is chosen as base we get indices of 100, 133·3 and 166·7. If the base is the second year the indices are 75, 100 and 125. In the event of the third year being chosen we have 60, 80 and 100.

It is clearly important that the base year be selected to provide a satisfactory standard of comparison. A *high* base year tends to detract from other years and a *low* one tends to make them appear higher. For this reason an average of several years is sometimes used to form an artificial base.

As time moves on the base may become unrealistic. This can occur if there has been a pronounced trend in the data so that, say, the base year output bears little relation to current quantities. This may be of small importance if we are prepared to put up with a very large (or small) index number. Conversion may be made to any other year as base, simply by dividing all the indices by the index of the new base year on the old scale, and multiplying by 100. In the example just given the indices 100, 133·3 and 166·7 can be converted to the second year as base by multiplying each by $\frac{100}{133 \cdot 3}$.

A more serious objection to index number comparisons over long periods of time is that the basic circumstances change. Even if we are dealing with what seems to be a single commodity it rarely remains exactly the same. Designs change and quality varies.

This applies especially if the index number relates to a number of items, such, for example, as those bought by the average person. In times of rapid progress the consumption patterns may change almost out of recognition in a decade.

## 5.2 INDEX NUMBERS OF AGGREGATES

Table 5.1 represents a common index number problem. There are five commodities labelled A to E. Their prices are shown under $P_0$ and the quantities purchased under $Q_o$. The suffix 0 is generally used to indicate the base year. The prices must relate to the units in which the quantities are expressed, though these may vary from item to item. A may be in pounds weight, B in gallons, C in yards etc., so that the sum of these may have no meaning.

Table 5.1

| Item | $P_o$ | $Q_o$ | $P_oQ_o$ | $P_n$ | $P_nQ_o$ |
|---|---|---|---|---|---|
| A | 25 | 4 | 100 | 26 | 104 |
| B | 3 | 40 | 120 | 4 | 160 |
| C | 36 | 10 | 360 | 42 | 420 |
| D | 90 | 6 | 540 | 87 | 522 |
| E | 44 | 20 | 880 | 54 | 1080 |
| | 198 | | 2000 | 213 | 2286 |

The total expenditure in the base year is found by multiplying the price by the quantity for each commodity, and then adding these products, to give $\Sigma P_oQ_o$.

If we have the prices for some other year (n), shown under $P_n$ we can multiply these by the old quantities to find the cost of maintaining the base year pattern of purchases, obviously $\Sigma P_nQ_o$.

The weighted index number for year n ($Y_n$), being a price index, can be denoted by $P_{on}$, and is $\frac{\Sigma P_nQ_o}{\Sigma P_oQ_o}\times 100=\frac{2286}{2000}\times 100=114{\cdot}3$.

An unweighted index can be obtained simply by adding the prices without weighting them ($\frac{213}{198}\times 100=108$, in the example). This may be very misleading, unless the quantities to which the prices relate are themselves roughly in proportion to the purchases.

Slight variations in weights make little difference to the result, and extreme accuracy is not needed.

## 5.3 PRICE RELATIVES

It is often useful to show the relative change compared with the base data of each item, or major group of items, included in a main index.

When price indices are involved such *price relatives* (as these subsidiary indices are called) may be combined to give the same result as the weighted index of all the items provided that the weights used are proportional to the amount *spent* on each commodity during the base year ($P_oQ_o$). The reader should reason out why the amount spent is used rather than the simple quantities bought, using algebraic notation.

The Retail Price Index is split up in this way. It has the advantage that it is easy to prepare an index relating to patterns of expenditure differing from the normal (see worked example).

Table 5.2 shows the price relatives for the previous data, each being $\frac{P_n}{P_o}\times 100$. The weights, as percentages, are calculated from $\frac{P_oQ_o}{2000}\times 100$.

The price relative and weight for each item are then multiplied and the products are totalled. The index is this sum divided by 100, i.e. $\frac{11430{\cdot}1}{100}=$ 114·3, as previously calculated directly.

The nature of the data entering into most index numbers make decimal point accuracy spurious.

Table 5.2

| Item | Price Relative | Weight | Weight× Price Relative |
|---|---|---|---|
| A | 104·0 | 5 | 520·0 |
| B | 133·3 | 6 | 799·8 |
| C | 116·7 | 18 | 2100·6 |
| D | 96·7 | 27 | 2610·9 |
| E | 122·7 | 44 | 5398·8 |
| | | 100 | 11430·1 |

## 5.4 QUANTITY (OR VOLUME) INDICES

Apart from simple cases where there is a single item, as in the first section of this chapter, there are two approaches to forming a quantity index where amounts cannot be added together directly, because the units or the commodities themselves differ. We cannot add pounds to yards nor can we add tons of steel to tons of coal.

The first approach is to add the values of the items, and then to remove any effects of price changes, by dividing by a price index for the year concerned, and multiplying by 100.

The process of dividing data by an appropriate index is known as *deflating* the data. An example would be dividing an index of cash income by a price index to arrive at an index of real income.

The second approach is to reverse the thinking of the price index. Suppose we weight the quantities for the years concerned with the prices of the base year, then the resulting total is a measure of how much the total of these quantities has changed.

The quantity index will be $Q_{on}=\frac{\Sigma P_oQ_n}{\Sigma P_oQ_o}\times 100$. Suppose that, in our previous example, we have the additional information that the commodities for $Y_n$ are represented by quantities ($Q_n$) of 7, 45, 20, 8 and 10 respectively. The terms $P_oQ_n$ become 175, 135, 720, 720 and 440, respectively, totalling 2190. The quantity index for $Y_n$, based on $Y_o$, is thus $\frac{2190}{2000}\times 100=109{\cdot}5$.

## 5.5 LASPEYRE, PAASCHE ETC. AND REVERSAL TESTS

The type of index explained so far is referred to as a *Laspeyre* type, its characteristic being that the weighting applicable to the base year is used. The price index is the cost, in $Y_n$, of buying the same commodities as in the base year.

If we work backwards, comparing the actual expenditure in $Y_n$ with what it would have cost to buy the nth year's products in the base year, we have a *Paasche* type of index (often shown as $P_{on}$). A problem is that the weights change as we move forward, but it may be argued that the comparison it makes is more relevant to the developing circumstances. The index of price is $\frac{\Sigma P_n Q_n}{\Sigma P_o Q_n} \times 100$, and the quantity index is $\frac{\Sigma P_n Q_n}{\Sigma P_n Q_o} \times 100$.

Another approach is to compromise between base year and nth year weights. The *Marshall-Edgeworth* index uses the sum of the quantities for the two years as weights.

*Fisher's 'Ideal'* index uses the square root of the product of the Laspeyre and Paasche indices. It is difficult to interpret but satisfies the theoretical requirements of both the *time reversal* and *factor reversal* tests.

The former implies that, if we regard $Y_n$ as the base year and calculate the index for $Y_o$, it is the reciprocal of the index for $Y_n$ based on $Y_o$. The two indices are, for $Y_n$ on $Y_o$, $\sqrt{\frac{\Sigma P_n Q_o}{\Sigma P_o Q_o} \times \frac{\Sigma P_n Q_n}{\Sigma P_o Q_n}}$, and, for $Y_o$ on $Y_n$ $\sqrt{\frac{\Sigma P_o Q_n}{\Sigma P_n Q_n} \times \frac{\Sigma P_o Q_o}{\Sigma P_n Q_o}}$, and their product is unity.

The latter requires that, if we multiply a price index and a quantity index together, the quantity index being produced by reversing the factors, that is substituting P for Q and vice versa in the price index, while keeping the suffixes the same, then the product will be an index of the *relative expenditures* at the two dates. Thus the Fisher quantity index becomes $\sqrt{\frac{\Sigma Q_n P_o}{\Sigma Q_o P_o} \times \frac{\Sigma Q_n P_n}{\Sigma Q_o P_n}}$. When multiplied by the price index previously shown we have $\sqrt{\left(\frac{\Sigma P_n Q_o}{\Sigma P_o Q_o} \times \frac{\Sigma P_n Q_n}{\Sigma P_o Q_n}\right) \times \left(\frac{\Sigma Q_n P_o}{\Sigma Q_o P_o} \times \frac{\Sigma Q_n P_n}{\Sigma Q_o P_n}\right)}$, which cancels down to $\frac{\Sigma P_n Q_n}{\Sigma P_o Q_o}$, obviously an expenditure index, sometimes called a *value relative*.

If we take the data previously used the sums of the products of prices and quantities are $\Sigma P_o Q_o = 2000$, $\Sigma P_n Q_o = 2286$, $\Sigma P_o Q_n = 2190$ and $\Sigma P_n Q_n = 2438$.

The Paasche price index for $Y_n$ is thus $P'_{on} = \frac{2438}{2190} \times 100 = 111{\cdot}2$.

The Fisher price index is $\sqrt{\frac{2286}{2000} \times \frac{2438}{2190}} = \sqrt{114{\cdot}3 \times 111{\cdot}2} = 112{\cdot}7$.

The reader should similarly calculate the Paasche and Fisher quantity indices, and also the Marshall-Edgeworth indices for prices and quantities (using weight of $Q_o + Q_n$ and $P_o + P_n$ respectively).

Sometimes the *geometric mean* is used in index calculation. This implies that we multiply together the price relatives, each raised to the power denoted by the weights we are using, and then take the root equal to the sum of the weights. This will generally be done by logarithms.

The geometric mean satisfies the time reversal test.

Students should be prepared to meet different symbols used for index numbers from those used here. The letter I is often used for price indices (with J for quantity indices). W may indicate a weighted index compared with S, a simple index.

## 5.6 LINK AND CHAIN RELATIVES

With *link relatives* the prices or volumes of each year in a series are expressed as percentages or ratios of the preceding year. The base thus moves forward, and the indices are designated as $P_{01}$, $P_{12}$, $P_{23}$ etc. In the example at the start of this chapter the index for the second year, based on the first year, is $\frac{120}{90}\times 100=133{\cdot}3$; that for the third year, based on the second year, is $\frac{150}{120}\times 100=125$.

If we wish to relate a chosen year to a base year other than the preceding year, we can do this by multiplying together all the link relatives involved between the two years. This is referred to as *chaining* the relatives to a fixed base.

In the example just given the index for the third year, based on the first would be $\frac{120}{90}\times\frac{150}{120}\times 100$, or, using the indices, $\frac{133{\cdot}3}{100}\times\frac{125}{100}\times 100=166{\cdot}7$, the same as was calculated directly.

An advantage of this system is that the weights used can be gradually altered to meet changing conditions.

## 5.7 PUBLISHED INDEX NUMBERS

Below are given brief details of three of the best-known index numbers.

a. *Index of Retail Prices*

There used to be a Cost of Living Index, based on 1914 and relating to working class living standards. This was replaced in 1947 by an Interim Index of Retail Prices, which gave way in 1956 to the Index of Retail Prices, later rebased on 1962. The aim of the present index is 'to enable price changes to be measured over the whole field of goods and services over which households distribute their expenditure'. The weighting of the main groups (which themselves are divided into sections) is amended each January, reflecting the findings of the *Family Expenditure Survey*, which examines the expenditure patterns of a sample of households (see section 5.9). The index is compiled by the Department of Employment and Productivity.

b. *Wholesale Price Index*

The first official index was published in 1903, replaced in 1921 and again in 1935 by the Board of Trade Index Number of Wholesale Prices, based on 1930 and using a geometric mean.

In 1951 a new series of index numbers was started for different classes of commodities, avoiding some of the pitfalls of a composite index. The most recent basing is on 1963.

c. *Index of Industrial Production*

An Interim Index was published in 1948, to be replaced in 1952 by the Index of Industrial Production. This has been rebased several times, the latest being 1963. The purpose is 'to reflect changes in the volume of industrial production in the United Kingdom'. Agriculture, forestry, fishing, distribution, transport, finance and all services are excluded. It is a weighted arithmetic average of quantity relatives, these generally being measured by changes in the outputs of the industries, while the weights relate to the net outputs as shown by the Census of Production, now to be held every five years. Over 800 series are incorporated in this manner.

## 5.8 RATIOS

A *ratio* is the relationship of a quantity or size to another of the same kind. It is often a useful device for making comparisons when absolute amounts are misleading. Thus, a comparison of the profits of two firms is not very revealing unless we relate them to some factor, such as the capital employed.

Such *management ratios* have become accepted and are usually regarded as appropriate to the field being studied, e.g. accounting or personnel. Examples are:

a. Rate of labour turnover or separation rate

$$=\frac{\text{number of workers leaving in the period}}{\text{average number on the payroll}}\times 100.$$

b. $\dfrac{\text{Profit}}{\text{Total Assets}}, \dfrac{\text{Profit}}{\text{Net Worth}}$ and $\dfrac{\text{Profit}}{\text{Sales}}$.

c. $\dfrac{\text{Sales}}{\text{Stock}}, \quad \dfrac{\text{Sales}}{\text{Cash}}, \quad \dfrac{\text{Sales}}{\text{Debtors}}, \quad \dfrac{\text{Sales}}{\text{Current Assets}}, \quad \dfrac{\text{Sales}}{\text{Fixed Assets}}.$

d. $\dfrac{\text{Production Cost}}{\text{Sales}}, \quad \dfrac{\text{Direct Labour Cost}}{\text{Sales}}, \quad \dfrac{\text{Marketing Cost}}{\text{Sales}}$ etc.

e. $\dfrac{\text{Current Assets}}{\text{Current Liabilities}}.$

Ratios have long been employed in branches of social statistics. In *vital statistics* (concerned with human beings) we have ratios such as:

a. Crude birth rate $=\dfrac{\text{number of births}}{\text{population}}\times 1{,}000$, and the crude death rate on similar lines. A rate per thousand of the population is used since it gives a convenient figure.

b. Standardised birth rates (and death rates) eliminate the effects of differences in age and sex structure between different populations by selecting a standard distribution of age and sex and calculating a sort of index number, combining the actual rates in each group in each population.

c. Specific birth (and death) rates are related to particular age (or other) groups. In the case of the birth rate per thousand women aged 15 to 44 (called the *general fertility rate*) the attempt is to relate births only to those members of the community generally capable of child-bearing. The *infant mortality rate* was devised to deal with children under one year old.

d. Net reproduction rates attempt to show the trend in future population based on the numbers in which women will reproduce themselves. The birth rates to different age groups and also the death rates are taken into account, and the ratio of the number of female children who will survive as adults to the present number of adult females is calculated. Unity implies a level population. Greater (less) than unity implies an increase (decrease) in population. If the effect of mortality is ignored we have the *gross reproduction rate*.

e. $$\text{Morbidity rate} = \frac{\text{Number of cases of a disease}}{\text{Population}} \times 1{,}000.$$

## 5.9 WORKED EXAMPLE

*Question*

Table 5.3 shows the main divisions of the Department of Employment and Productivity Index of Retail Prices, with the weightings and percentage relatives for a certain month. (The base is at 16th January, 1962).

Table 5.3. *Index of Retail Prices*

| | Item | Weight (W) | Price Relative (R) | $\frac{W \times R}{1000}$ |
|---|---|---|---|---|
| i | Food | 263 | 125·4 | 33·0 |
| ii | Alcoholic drink | 63 | 132·7 | 8·4 |
| iii | Tobacco | 66 | 134·8 | 8·9 |
| iv | Housing | 121 | 143·6 | 17·4 |
| v | Fuel and light | 62 | 138·2 | 8·6 |
| vi | Durable household goods | 59 | 115·4 | 6·8 |
| vii | Clothing and footwear | 89 | 114·7 | 10·2 |
| viii | Transport and vehicles | 120 | 122·5 | 14·7 |
| ix | Miscellaneous goods | 60 | 128·0 | 7·7 |
| x | Services | 56 | 137·7 | 7·7 |
| xi | Meals bought out | 41 | 130·3 | 5·3 |
| | | 1000 | | 128·7 |

*a.* Calculate the index including all these items.

*b.* Assuming that your firm sells alcoholic drink and tobacco prepare an index number relating to these and compare their price changes with those of food.

*Answer*

*a.* The price relatives have been multiplied by the appropriate weights and the products divided by 1000, to economise on figures. The total then gives the required index.

Index = 128·7.

*b.* An index for alcoholic drink and tobacco can be formed as follows:

| *Item* | W | R | WR |
|---|---|---|---|
| Drink | 63 | 132·7 | 8360 |
| Tobacco | 66 | 134·8 | 8897 |
| | 129 | | 17257 |

Index is $\frac{17257}{129}$ =133·8, which has to be compared with the index for food of 125·4, showing that the items in question have risen more than food prices by approximately 8 points (133·8—125·4) compared with the base year.

[The reader should note that the weights of this index are revised each January and should not assume the weights in the question apply currently.]

## 5.10 PRACTICE QUESTIONS

1. A firm divides its materials into four main groups and tries to estimate the over-all effect of price changes by producing an index number weighted according to the tonnages used in 1965. Calculate this index for 1969 from the following information, and comment on the result.

| Group | 1965 | | 1969 | |
|---|---|---|---|---|
| | *Tons* | *Price* (£) | *Tons* | *Price* (£) |
| A | 50 | 85 | 45 | 116 |
| B | 120 | 34 | 185 | 42 |
| C | 35 | 10 | 68 | 15 |
| D | 210 | 48 | 250 | 50 |

2. Produce a Paasche type price index for the data in question 1. Comment on the relative merits of this compared with a base-weighted index from management's point of view.

3. Convert the data in question 1 into the form of price relatives and combine these into an index based on 1965. Account for the weighting you have used.

4. Produce a quantity index for the data in question 1 assuming that A is in tons, B is in square yards, C is in gallons and D is in units. Discuss the interpretation of such an index.

5. The values of a certain index number over eight years are 102, 130, 125, 120, 135, 140, 160, 155. Calculate
   *a.* The index for year 8 based on year 4.
   *b.* The index for year 3 based on year 7.
   
   Comment on the validity of this process.

6. Discuss the merits of producing an index of retail prices for different areas and for different social groups.

# 6

# Dispersion and Skewness

## 6.1 MEANING OF DISPERSION

While the average is probably the most useful way of describing data briefly it can obscure important aspects of the detailed figures. It indicates the central tendency of data and ignores the general pattern we discussed in chapter 2.

Dispersion is concerned with how the data are spread out. The various measures of dispersion attempt to concentrate this information in a single figure.

The chief measures of dispersion are:

a. Range
b. Mean deviation
c. Quartile deviation
d. Standard deviation, which is by far the most important.

Table 6.1. *Number of units sold by four salesmen over an eight-week period*

| Week | Salesman | | | |
|---|---|---|---|---|
| | A | B | C | D |
| 1 | 38 | 28 | 0 | 29 |
| 2 | 19 | 27 | 70 | 28 |
| 3 | 35 | 30 | 0 | 32 |
| 4 | 15 | 25 | 0 | 14 |
| 5 | 21 | 25 | 74 | 30 |
| 6 | 32 | 28 | 0 | 31 |
| 7 | 34 | 26 | 72 | 30 |
| 8 | 22 | 27 | 0 | 30 |
| | 216 | 216 | 216 | 224 |

The nature of the problem is clarified by a study of table 6.1. It will be seen that salesmen A, B and C have the same average performance but the variability from week to week differs greatly. D has the best average performance but his sales for week 4 are lower than the sales of A or B for any individual week.

The study of variability is of great importance. To start with we are concerned with measuring it satisfactorily to afford a basis of comparison between different sets of figures. For more advanced work the analysis of this variability and its assignment to its causes provides one of the most powerful of statistical tools.

## 6.2 RANGE

This affords the most obvious, the most easily arrived at, and probably the most easily understood of all the measures of dispersion.

It is simply the difference between the largest and the smallest items.

In the case of the sales figures the results for A, B, C and D respectively are 23, 5, 74 and 18 units. With such simple data these ranges give a fairly good indication of the variability of the salesmen.

The snag with the range is that it depends on only two items which, by definition, are extremes. In the case of salesman D, if the single low figure of 14 were eliminated, the range would become 4 instead of 18. The range gives no indication of the degree of concentration of the mass of the data about the average. With grouped data the range may not be obtainable.

## 6.3 MEAN DEVIATION

The obvious solution to the deficiencies of the range is to calculate some form of average of the amount by which each item differs from the mean that has been chosen to represent the data.

Table 6.2. *Deviations of the sales figures from the average of each man*

| Week | Salesman | | | |
|---|---|---|---|---|
| | A | B | C | D |
| 1 | 11 | 1 | −27 | 1 |
| 2 | −8 | 0 | 43 | 0 |
| 3 | 8 | 3 | −27 | 4 |
| 4 | −12 | −2 | −27 | −14 |
| 5 | −6 | −2 | 47 | 2 |
| 6 | 5 | 1 | −27 | 3 |
| 7 | 7 | −1 | 45 | 2 |
| 8 | −5 | 0 | −27 | 2 |
| | −31 | −5 | −135 | −14 |
| | +31 | +5 | +135 | +14 |
| Total (ignoring sign) | 62 | 10 | 270 | 28 |

Table 6.2 shows the calculation of the difference of each item from the average of its respective salesman. The total of these differences should always be zero, since this is what the arithmetic mean implies.

If we ignore the minus signs the totals are as shown at the bottom of the table. Division of each of these by the number of items included (8) gives values of 7·75, 1·25, 33·75 and 3·50 units for A, B, C and D respectively.

This average of the deviations from the average is termed the *mean deviation.*

Table 6.3 shows the calculation of the mean deviation for the bonus data previously used. With grouped data such as this we have to multiply each deviation by the appropriate frequency and divide the total by the sum of the frequencies. The mean deviation is $\frac{87{\cdot}44}{150}=£0{\cdot}58$.

The mean deviation is easy to calculate and readily understood and is of use if we require only a measure of dispersion. If the measure is to be used as the basis of further calculations and judgments it is best to use the standard deviation (section 6.5.1).

Table 6.3. *Calculation of mean deviation for bonus data*

| Bonus mid-point | Frequency (f) | Deviation from 5·42 (d) | fd |
|---|---|---|---|
| 3·5 | 6 | −1·92 | −11·52 |
| 4·5 | 35 | −0·92 | −32·20 |
| 5·5 | 75 | +0·08 | +6·00 |
| 6·5 | 33 | +1·08 | +35·64 |
| 7·5 | 1 | +2·08 | +2·08 |
| | 150 | | +43·72 |
| | | | −43·72 |
| | | | 87·44 |

## 6.4 QUARTILE DEVIATION

This measure of dispersion makes use of the partition values (section 3.3.4) and is especially appropriate when the arithmetic mean has not been calculated and is not available for further working.

The quartile deviation takes half the difference between the upper and lower quartiles as an indication of the dispersion of the data. For this reason it is sometimes called the *semi-interquartile range.*

The calculation for the bonus data (using the ascending frequencies of table 2.7) is as follows.

1st (or lower) quartile ($Q_1$) is the $\frac{150}{4}=37\frac{1}{2}$th item, i.e. $Q_1=4+\frac{37{\cdot}5-6}{35}=4{\cdot}9$.

3rd (or upper) quartile ($Q_3$) is the $150\times\frac{3}{4}=112\frac{1}{2}$th item, i.e. $Q_3=5+\frac{112{\cdot}5-41}{75}$ $=5{\cdot}95$.

The quartile deviation is thus $\frac{Q_3-Q_1}{2}=\frac{5{\cdot}95-4{\cdot}9}{2}=£0{\cdot}525$.

## 6.5.1 STANDARD DEVIATION

This is by far the most important of the measures of dispersion and an appreciation of its calculation and properties is essential to anybody wishing to progress with statistics beyond the most elementary level.

The approach is similar to that of the mean deviation. Instead, however, of ignoring the minus signs it legitimately makes them positive by squaring all the deviations. These squared values are added and their average found by dividing by the number of items involved. Finally the square root of the result is calculated and is called the *standard deviation.* It is usually denoted by the small Greek sigma ($\sigma$), though, if we are using a sample of the data, the letter $s$ is used.

The formula is $\sigma=\sqrt{\frac{\Sigma(x-\mu)^2}{N}}$, where each item occurs singly with a total of N items, or $\sigma=\sqrt{\frac{\Sigma f(x-\mu)^2}{\Sigma f}}$, where the data are grouped, the frequency of each group being denoted by f.

Table 6.4 shows the calculation of the standard deviation for the sales data. The deviations of table 6.2 have been squared and totalled for each column. The results for A, B, C and D are found from $\sqrt{\frac{528}{8}}$, $\sqrt{\frac{20}{8}}$ etc., giving standard deviations of 8·1, 1·6, 34·9 and 5·4 units respectively.

Table 6.4. *Calculation of standard deviation for sales data*

| Week | Squared Deviations A | B | C | D |
|---|---|---|---|---|
| 1 | 121 | 1 | 729 | 1 |
| 2 | 64 | 0 | 1849 | 0 |
| 3 | 64 | 9 | 729 | 16 |
| 4 | 144 | 4 | 729 | 196 |
| 5 | 36 | 4 | 2209 | 4 |
| 6 | 25 | 1 | 729 | 9 |
| 7 | 49 | 1 | 2025 | 4 |
| 8 | 25 | 0 | 729 | 4 |
| | 528 | 20 | 9728 | 234 |

The straightforward extension of this method to grouped data is shown in table 6.5. It is assumed that the mean has already been calculated (as in chapter 3). The deviations from this mean are shown in the column headed d. These deviations are squared, multiplied in each case by the appropriate value of f, and these products are totalled. The standard deviation is then $\sqrt{\frac{94{\cdot}99}{150}}=£0{\cdot}79$.

Table 6.5. *Straightforward calculation of standard deviation from bonus data*

| Bonus (x) | Frequency (f) | Deviation from 5·42 (d) | $d^2$ | $fd^2$ |
|---|---|---|---|---|
| 3·5 | 6 | −1·92 | 3·686 | 22·12 |
| 4·5 | 35 | −0·92 | 0·846 | 29·61 |
| 5·5 | 75 | 0·08 | 0·006 | 0·45 |
| 6·5 | 33 | 1·08 | 1·166 | 38·48 |
| 7·5 | 1 | 2·08 | 4·326 | 4·33 |
| | 150 | | | 94·99 |

This approach can be modified to avoid the awkward figures which usually occur when the deviations are calculated from the actual mean. This calculation by the short-cut method is explained in the next section.

### 6.5.2 SHORT-CUT METHODS

Table 6.6 shows the basic working for the calculation of the standard deviation for the output data previously used. The approach is an extension of that used to obtain the arithmetic mean, working from an assumed mean. In this case 95 has been chosen and deviations from this, in 10's, are shown in the column headed *d*. These deviations are multiplied by the frequency *f* to give the column headed fd. Further multiplication of this column by *d* gives the column headed $fd^2$.

Table 6.6. *Calculation of standard deviation of output data working from assumed mean*

| Output (x) | Frequency (f) | Deviation from 95 in 10's (d) | fd | $fd^2$ |
|---|---|---|---|---|
| 65 | 2 | −3 | −6 | 18 |
| 75 | 6 | −2 | −12 | 24 |
| 85 | 10 | −1 | −10 | 10 |
| 95 | 23 | 0 | 0 | 0 |
| 105 | 8 | 1 | 8 | 8 |
| 115 | 1 | 2 | 2 | 4 |
| | 50 | | −28 | 64 |
| | | | +10 | |
| | | | −18 | |

The algebraic sum of the fd column (−18), when divided by the total frequency (50) and multiplied by the working units (10) gives the adjustment needed to find the mean. The mean is thus $95-\frac{18}{50}\times 10=91\cdot 4$ tons.

To find the standard deviation we divide the total of the squares column (64) by the total frequency (50) and *deduct* from this the square of the adjustment used to find the mean (before it has been multiplied by the working units). The square root of the difference multiplied by the working units gives the required standard deviation, which is $\sigma=10\left(\sqrt{\frac{64}{50}-\left(\frac{18}{50}\right)^2}\right)=$ 10·7 tons.

With regard to the adjustment that is made when using this method it should be noted that, for the mean, whether this is added or subtracted depends entirely on the sign of the sum of the fd column, while, for the standard deviation, the square of this amount is always subtracted.

Notice, also, that it is best to calculate to the square root stage in terms of the working units and then to multiply this result by the working units to give the required result in terms of the original units.

We may take the assumed mean at any value we like. If we pick zero then the data remain the same. The adjustment becomes equal to the arithmetic mean. We square the original data and multiply by the frequency of each group. These products are added and the sum is divided by the total frequency. From this we subtract the square of the arithmetic mean. The result is the variance (see 6.5.6), of which we require the square root.

If we are dealing with a series of discrete numbers all we require are the totals of the numbers and of their squares. For the numbers 1 to 5 inclusive the total is 15 so that the average is 3, of which the square is 9. The total of the squares is 55 which, divided by the number of items (5), gives 11. The standard deviation of these numbers is thus $\sqrt{11-9}=\sqrt{2}$.

## 6.5.3 STANDARDISED DATA

It is sometimes convenient to *standardise* data so that different sets of data can be compared.

To apply this we require to know the arithmetic mean and the standard deviation. The mean is regarded as the centre of the data and each item is expressed as a plus or minus deviation from it. These deviations are divided by the standard deviation.

Table 6.7. *Data standardised in terms of the standard deviation*

| $x-\bar{x}$ | $\frac{x-\bar{x}}{\sigma}$ | $\frac{f}{\Sigma f}$ |
|---|---|---|
| −26·4 | −2·47 | 0·04 |
| −16·4 | −1·53 | 0·12 |
| −6·4 | −0·60 | 0·20 |
| +3·6 | 0·34 | 0·46 |
| +13·6 | 1·27 | 0·16 |
| +23·6 | 2·21 | 0·02 |
| | | 1·00 |

Table 6.7 shows this process applied to the output data. If we take an item such as the group centred on 115 the deviation from $\mu$ (91·4) is 115−91·4=23·6, which, divided by $\sigma$ (10·7) gives 2·2. Thus, instead of 115, we can write $91{\cdot}4+2{\cdot}2\sigma$ or $\mu+2{\cdot}2\sigma$.

The frequencies can also be standardised as percentages of the total, or the whole area under the histogram can be considered as unity and the frequencies designated as decimal fractions of this. This has been done in the last column of table 6.7.

This gives a mathematical concept of the data, which replaces the original units and frequencies and is of enormous value in the treatment of standard patterns that many distributions follow. This is developed in chapter 8.

## 6.5.4 DEGREES OF FREEDOM AND BESSEL'S CORRECTION

The data we use for the calculation of the standard deviation may generally be regarded as a sample drawn from a much larger population. The mean of this sample ($\bar{x}$) will tend to differ from the accurate mean of the population (shown as $\mu$). This will cause the standard deviation obtained from the sample to differ from what it would be if calculated from the population mean.

*Bessel's correction* allows for this. We multiply the value of the standard deviation, as already obtained, by $\sqrt{\frac{n}{n-1}}$. This is equivalent to using n−1 instead of n in the original formula for the standard deviation. Here *n* is the number of items in the sample, as distinct from *N*, the number in the population.

As n becomes larger the top and bottom of the correcting factor become nearly the same and the change involved is negligible.

The reduction of the number of items by one for calculation purposes is a special example of what is called the *degrees of freedom* of data. Briefly, if we have n items totalling a given amount, then we may fill in n—1 of the items freely. Having done this the last item is really fixed by the total.

Let us again consider the sales data of table 6.1.

The degrees of freedom associated with the 32 items considered separately are 32—1=31.

Each column has 8 items. The 4 column totals equal the total of all the items in the table. The degrees of freedom for each column are 8—1=7. Since there are 4 columns we multiply this by 4 to give 28. Associated with the 4 column totals are 4—1=3 degrees of freedom. The total of 28 and 3 is 31, the total degrees of freedom for the data.

If the data were added across to give row totals for sales each week we should have 8 rows with 3 degrees of freedom in each, giving 8×3=24 degrees of freedom, and 8 row totals, giving 8—1=7 degrees of freedom. Again the total is 31 degrees of freedom.

Degrees of freedom are of considerable importance in the more advanced study of variability which is dealt with in chapter 20.

## 6.5.5 SHEPPARD'S CORRECTION

This is a correction sometimes applied to allow for the error that arises from our assumption that, with grouped data, the centre of the group can be considered as representing the group.

The adjustment simply involves subtracting $\frac{I^2}{12}$ (where I is the group interval or width) immediately before taking the square root, when calculating the standard deviation.

This correction applies to continuous distributions that tail off to zero at each end. Since there can easily arise an over-correction by injudicious application of the correction the student is advised to ignore this adjustment for examination calculations, unless it is specifically required.

If the original data has been adjusted and the calculation done in terms of a working variable the value of I used must relate to this working variable.

## 6.5.6 VARIANCE

The square of the standard deviation is known as the *variance*. It is calculated in the same way as the standard deviation with the exception that we do not take the square root. If we are working in coded units we must multiply the variance in terms of these working units by the square of the working unit.

For the output data the variance would be calculated from $10^2\left(\frac{64}{50}-\left(\frac{18}{50}\right)^2\right)$.

The *analysis of variance*, considered in chapter 20, has many applications.

For the moment we may note that the variances of any number of samples from the same population may be added.

The estimate of the variance from each sample is multiplied by the number of items in its sample. These products from each sample are added and the sum is divided by the number of items in all the samples, less one for each sample included.

If there are three samples yielding variances of $s_1^2$, $s_2^2$ and $s_3^2$ and having $n_1$, $n_2$ and $n_3$ items, respectively, then the estimate of the population variance (usually shown as $\hat{\sigma}^2$) would be given by $\frac{n_1s_1^2+n_2s_2^2+n_3s_3^2}{n_1+n_2+n_3-3}$.

This provides another example of the use of degrees of freedom.

## 6.6 COEFFICIENTS OF DISPERSION

The measures of dispersion so far considered are expressed in the units of the data to which they refer. A standard deviation of £1 may refer to data averaging hundreds of pounds or only a few pounds. The relative deviation in the first case is clearly much less than in the second.

*Coefficients of dispersion* provide a relative measure. The particular measure of dispersion is divided by an appropriate average. This yields a measure which is a pure number and does not require the units to be stated.

Coefficients enable comparisons to be made of the dispersions of distributions expressed in different units. An example would arise if income distributions were being compared and the incomes were expressed in different currencies.

The chief coefficients of dispersion are:

a. *Coefficient of variation*, where the standard deviation is divided by the arithmetic mean, the quotient being subsequently multiplied by 100. This gives $\frac{100s}{\bar{x}}$.

b. A coefficient based on the mean deviation by dividing this either by the arithmetic mean or by the median.

c. *Coefficient of quartile deviation* found by dividing the quartile deviation by the median. An alternative is to divide by the average of the quartiles, which gives a result equivalent to $\frac{Q_3-Q_1}{Q_3+Q_1}$.

## 6.7 MOMENTS

The arithmetic mean and the variance are really special cases of what statisticians generally refer to as *moments*.

The concept of moments is familiar to scientists in connection with the balancing of forces about a fulcrum, taking account of the distances involved, and also in connection with moments of inertia.

If we consider the simple lever shown in diagram 6.1 it will be seen that this is in balance, since the moments tending to rotate it clockwise ($5\times1+1\times3+2\times5=18$ lb. ft.) are balanced by that tending to cause anti-clockwise rotation ($6\times3=18$ lb. ft.).

This is really equivalent to the distribution of values about the average. The distance from A is the value of the item, and its force is the equivalent of the frequency. Calculation of the average on this basis would give $(6\times0+5\times4+1\times6+2\times8)\div14=\frac{42}{14}=3$, which is the distance of the fulcrum F from A. The origin can be taken at any point, not necessarily at A.

The Greek letter $\mu$ (mu) is generally used to denote the moments, the study of which seems to be growing in importance. We have already used it to denote the population mean, which is really $\mu_1$.

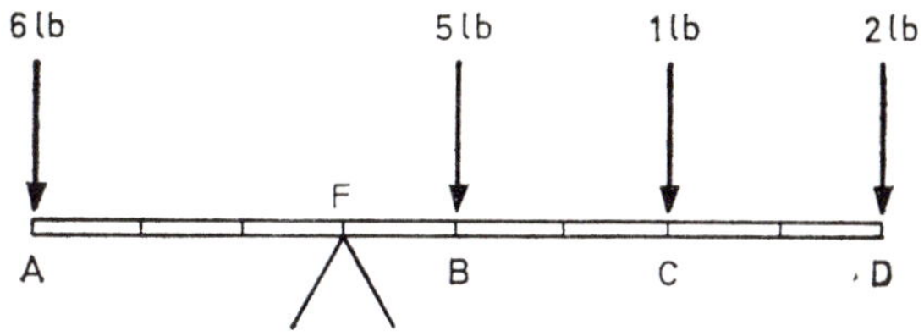


Diagram 6.1. *Mechanical analogy of moments*

The first moment is calculated about the origin (A) and is equivalent to the arithmetic mean.

$$\text{1st moment} = \mu_1 = \frac{\Sigma f(x-A)}{\Sigma f}$$

The second and higher moments are calculated about the mean as origin and are as follows:

$$\text{2nd moment} = \mu_2 = \frac{\Sigma f(x-\bar{x})^2}{\Sigma f}$$

$$\text{3rd moment} = \mu_3 = \frac{\Sigma f(x-\bar{x})^3}{\Sigma f}$$

$$\text{4th moment} = \mu_4 = \frac{\Sigma f(x-\bar{x})^4}{\Sigma f}, \text{ etc.}$$

It will be seen that the even-numbered moments must be positive whereas the odd-numbered moments can be negative.

Examples of the application of moments are afforded by the measures of skewness and kurtosis in the sections which follow.

In a more mathematical treatment of statistics we should probably study *moment generating functions*. These are functions of a variable which, expanded as a power series, give the moments of the frequency distribution as coefficients of the powers.

## 6.8 SKEWNESS

The third moment can be used as a measure of *skewness*, which is the tendency of a distribution to depart from symmetry.

If a distribution is symmetrical positive and negative deviations cancel out in moments of odd order, so that the third moment is zero. Any balance of the histogram to right or left will be reflected in the third moment, which will become positive or negative respectively.

The size of the third moment is also affected by the dispersion. The general effect of this can be eliminated by dividing by the second moment, adjusted to make it of the same order. This is equivalent to dividing by the cube of the standard deviation.

A measure of skewness incorporating these ideas is given by $\frac{\mu_3}{\mu_2^{3/2}} = \frac{\mu_3}{\sigma^3} = \frac{\Sigma f(x-\bar{x})^3}{\Sigma f} \div \sqrt{\left(\frac{\Sigma f(x-\bar{x})^2}{\Sigma f}\right)^3}$.

This may prove difficult to calculate.

Alternative measure of skewness are:

a. $\frac{\bar{x}-\text{mode}}{\sigma}$, and

b. $(Q_3+Q_1-2\times\text{median})\div(2\times\text{quartile deviation})$.

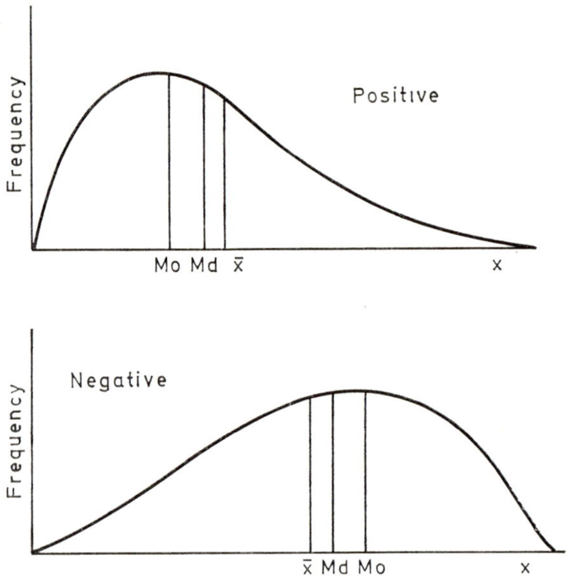


Diagram 6.2. *Skewness*, positive and negative, showing the displacement of the mean ($\bar{x}$), median (Md) and mode (Mo)

The first of these relies on the fact that, in a skew distribution, mode, median and mean no longer coincide (diagram 6.2). The distance separating the mode from the mean is taken as an indication of the degree of skewness. Again this is corrected for dispersion by dividing by the standard deviation.

The second formula involves comparison of the distance between the median and the upper quartile, and the median and the lower quartile, corrected for dispersion by dividing by the distance between the upper and lower quartiles. The formula is equivalent to $\frac{(Q_3-\text{median})-(\text{median}-Q_1)}{Q_3-Q_1}$.

The three measures of skewness give different results when applied to the same data, but they share the characteristic that perfect symmetry gives a result of zero, while a histogram pulled to the left gives a negative value, and one pulled to the right a positive value.

## 6.9 KURTOSIS

The fourth moment can be used as a measure of *kurtosis*, which is an indication of whether a frequency distribution tends to be peaked at the centre or flat compared with the normal curve (see section 8.1).

The measure is $\frac{\mu_4}{\mu_2^2}$ or $\frac{\mu_4}{\sigma^4}$. For the normal curve the value of this is 3, which provides the basis for a slightly modified form of measure, $\frac{\mu_4}{\mu_2^2}-3$, which will be zero for the normal distribution, positive for a highly peaked distribution, and negative for a relatively flat distribution.

## 6.10 THE LORENZ CURVE

This is a special form of curve designed to show the degree to which some factor is equally or unequally distributed among a population. It could be used for income or capital distribution in a community or for the sizes of the shareholdings in a company.

Each axis of the graph shows a cumulative percentage. One relates to the frequencies and the other to the factor being considered.

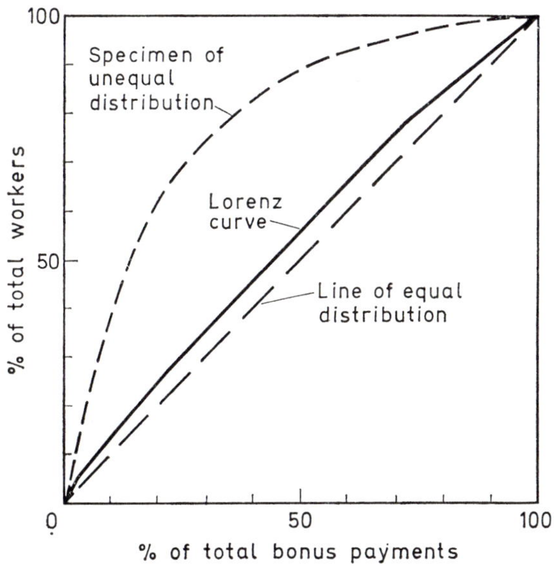


Diagram 6.3. *Lorenz curve* for bonus data

There are many ways of drawing the Lorenz curve. That illustrated (diagram 6.3) is probably the most appropriate for business use. The normal convention of showing frequencies vertically is maintained and there is no reversal of one of the scales, as is required by some versions of the curve.

Table 6.8 gives the calculations for the bonus data. Here we have to estimate the total payment to each group by multiplying the centre of the group by the group frequency in each case. Often data is presented in such a way that this is unnecessary, the total appropriate to each group being given. This would be the case if we were preparing the trade union data of table 3.9 for this form of display.

Table 6.8. *Bonus data prepared for Lorenz curve*

| Bonus Group Centre | No. of Items | Estimate of Amount Involved | Amount as % of Total: Group | Amount as % of Total: Cumulative | Cumulative Frequency % |
|---|---|---|---|---|---|
| 3·5 | 6 | 21·0 | 2·6 | 2·6 | 4·0 |
| 4·5 | 35 | 157·5 | 19·4 | 22·0 | 27·3 |
| 5·5 | 75 | 412·5 | 50·7 | 72·7 | 77·7 |
| 6·5 | 33 | 214·5 | 26·4 | 99·1 | 99·3 |
| 7·5 | 1 | 7·5 | 0·9 | 100·0 | 100·0 |
| | 150 | 813·0 | 100·0 | | |

The third column is the product of the first two columns.
The final column has already been calculated in table 2.7.
The two final columns are the basis of the Lorenz curve.

Notice that the original groupings do not appear in the curve. The values shown horizontally are based on the accumulated total values up to each group. For the bonus data these are sums of money. For the trade union data they would be the numbers of members in each of the stated sizes of union; the values shown vertically in this case would be based on the number of unions in each group.

With a fairly even distribution the curve tends to be a straight line connecting the origin of the graph with the 100% point of each scale. As the distribution becomes more uneven the curve tends to be pulled into the top left corner.

## 6.11 WORKED EXAMPLE

*Question*

A large retail chain of shops tested self-service operation in one of its shops prior to converting the rest of the chain. One point at issue was the effect of this selling method on the value of purchases of each customer. Records were kept showing how much each customer spent before and after conversion. Calculate and compare the means and standard deviations for these two distributions and comment on them.

| value (£) up to | —0·5 | —1·0 | —1·5 | —2·0 | —2·5 | —3·0 | —3·5 | —4·0 | over 4·0 |
|---|---|---|---|---|---|---|---|---|---|
| % before | 15·2 | 20·4 | 19·6 | 15·2 | 13·1 | 5·1 | 6·5 | 3·4 | 1·5 |
| % after | 11·3 | 16·8 | 23·4 | 18·2 | 14·1 | 6·9 | 4·7 | 2·6 | 2·0 |

*Answer*

The calculation of the mean and standard deviation, working from an assumed mean of 1·75 (the centre of group 1·5—2·0) in units of 0·5 is shown in table 6.9.

$$\text{Mean purchases (before)} \quad = \quad 1{\cdot}75-\frac{42{\cdot}1}{100}\times 0{\cdot}5=1{\cdot}54$$

Mean purchases (after) $= 1{\cdot}75 - \frac{28{\cdot}5}{100} \times 0{\cdot}5 = 1{\cdot}61$

Standard devn (before) $= \sqrt{\frac{421{\cdot}9}{100} - \left(\frac{42}{100}\right)^2} \times 0{\cdot}5 = 1{\cdot}01$

Standard devn (after) $= \sqrt{\frac{367{\cdot}9}{100} - \left(\frac{28{\cdot}5}{100}\right)^2} \times 0{\cdot}5 = 0{\cdot}95.$

The average has increased by $1{\cdot}61 - 1{\cdot}54 = 0{\cdot}07$, which seems a small amount. As a percentage, however, this comes to $\frac{0{\cdot}07}{1{\cdot}54} \times 100 = 4{\cdot}5\%$, which might be an appreciable addition to the total turnover.

Table 6.9

| Dev. from 1·75 in 0·5's (D) | $f_1$ | $f_1D$ | $f_1D^2$ | $f_2$ | $f_2D$ | $f_2D^2$ |
|---|---|---|---|---|---|---|
| −3 | 15·2 | −45·6 | 136·8 | 11·3 | −33·9 | 101·7 |
| −2 | 20·4 | −40·8 | 81·6 | 16·8 | −33·6 | 67·2 |
| −1 | 19·6 | −19·6 | 19·6 | 23·4 | −23·4 | 23·4 |
| 0 | 15·2 | 0 | 0 | 18·2 | 0 | 0 |
| 1 | 13·1 | 13·1 | 13·1 | 14·1 | 14·1 | 14·1 |
| 2 | 5·1 | 10·2 | 20·4 | 6·9 | 13·8 | 27·6 |
| 3 | 6·5 | 19·5 | 58·5 | 4·7 | 14·1 | 42·3 |
| 4 | 3·4 | 13·6 | 54·4 | 2·6 | 10·4 | 41·6 |
| 5 | 1·5 | 7·5 | 37·5 | 2·0 | 10·0 | 50·0 |
| | 100·0 | −106·0<br>+63·9 | 421·9 | 100·0 | −90·9<br>+62·4 | 367·9 |
| | | −42·1 | | | −28·5 | |

There seems to have been no radical change in the dispersion of the data about these means as judged by the standard deviations. There is a slight reduction in the spread.

[This problem is discussed further, from the point of view of sampling, at 9.5.]

The coefficients of variation are:

a. before, $\frac{1{\cdot}01}{1{\cdot}54} \times 100 = 66$

b. after, $\frac{0{\cdot}95}{1{\cdot}61} \times 100 = 59$

which confirms this reduction, when the amount relative to the average is considered.

## 6.12 PRACTICE QUESTIONS

1. Critically examine the range as a measure of dispersion.
2. Calculate the standard deviation for the following data:

| | | | | | | | | | |
|---|---|---|---|---|---|---|---|---|---|
| x | 0—20 | —30 | —40 | —50 | —60 | —70 | —80 | —90 | —100 |
| f | 2 | 5 | 8 | 15 | 27 | 35 | 20 | 5 | 3 |

3. Calculate a measure of dispersion for the data of qu. 2, based on the quartiles. In what circumstances might such a measure be used instead of the standard deviation?

4. Use a short method to calculate the variance of the ten consecutive even numbers, starting at 2.

5. Two samples containing 30 and 40 items, respectively, are taken from the same population, and the standard deviations are found to be 7.0 and 7.3. Combine these to estimate the standard deviation of the population.

6. Calculate the standard deviations of the outputs of the two machines in table 3.4.

7. Calculate the mean and standard deviation of the costs of the five items of table 3.6.

8. Since almost all the items in the case of a normally distributed variable will be found within $3\sigma$ of each side of the mean why can we not just divide the range by 6 to obtain the standard deviation?

9. Explain what the effects on the standard deviation of qu. 2 would be of *a*. doubling all the frequencies, *b*. doubling all the x values, *c*. increasing all the x values by 40.

# 7

# Probability

## 7.1 THE MEANING OF PROBABILITY

We often speak of there being, say, a *fifty-fifty* chance of something happening. This is an expression of our judgment of how likely the event is to occur, presumably relating to something such as the chance of a head when a coin is tossed.

Probability may be assessed in two ways, either *empirically* (which involves observing how frequently the event happens in a large number of trials), or *theoretically* (or *deductively*) from the logic of the situation (there is no reason to suppose a head will occur more frequently than a tail in an evenly balanced coin). The latter is often called the *a priori* approach, implying we work from established abstract ideas.

An empirical study of coin tossing might show that from a sample of 10 there were 6 heads, indicating a probability of 60% (more usually shown as 0·6). A sample of 100 might show 52 heads, indicating a probability of 0·52. With very large samples we should, presumably, approach very closely the probability of 0·5, which is what we have logically deduced as the probability of a head on any toss of the coin.

The probability of something happening (a *success*) is often indicated by $p$, and of the event's not happening (a *failure*) by $q$. Since the event must either happen or not happen $p+q=1$, unity being the expression of *certainty*.

The *expectation* is the probability of something happening multiplied by its value. If 40 people enter a lottery, of which the outcome is left entirely to chance, the prize being £1, then the probability for each of winning is $\frac{1}{40}$, and each has an expectation of £1$\times\frac{1}{40}$. If we have a batch of 1000 items from a process that produces 2% defective items, the expectation of the number of defective items is $100\times\frac{2}{100}=20$.

Events to which probabilities relate may be *independent* (one does not depend on the other) or *dependent* (where if the first event has occurred it affects the probability of the second).

Tossing coins and throwing dice are good examples of independent events. Whether we have heads or tails the first time does not affect the probability of a head the second throw. A *run* of heads still leaves the chance of the next event as 0·5. It is the *gambler's fallacy* to assume otherwise. Only in the long run should the number of heads and tails even out.

Dealing cards or drawing lots without replacement are examples of dependent events. If an ace has been dealt, as first card, from a pack of 52 cards, then there are 3 aces left from 51 cards, so the chance of the second card's being an ace is $\frac{3}{51}$, instead of $\frac{4}{52}$, as for the first card. If the first card is not an ace then the chance of the second card's being an ace is $\frac{4}{51}$.

## 7.2 ADDITION AND MULTIPLICATION LAWS

The *addition* of probabilities is used when we are interested in two or more ways in which an event can occur. If we have a six-sided die each face has a probability of $\frac{1}{6}$. The probability of throwing either 1 or 2 is the sum of their respective probabilities, $\frac{1}{6}+\frac{1}{6}=\frac{1}{3}$. This may be considered as dividing the ways the event can occur into two groups, the first of which, containing 2 of the 6 ways, constitutes success, and the second, containing 4 of the 6 ways, is counted a failure.

The probability of dealing either an ace or a court card from a 52 card pack would be $\frac{4}{52}+\frac{12}{52}=\frac{4}{13}$, since there are 4 aces and 12 court cards.

It should be noted that the addition of probabilities applies to a single event and the sum of probabilities cannot exceed 1, which includes all the possibilities and implies certainty.

If we desire to calculate the probability of two or more events happening together or in succession then the *multiplication* of probabilities is involved.

The probability of throwing two sixes, either with separate dice at the same time, or with the same die in successive throws, is $\frac{1}{6}\times\frac{1}{6}=\frac{1}{36}$.

Suppose there is a football pool with six games to forecast. What are the chances of success if we fill in a line at random? Since each match can be won, drawn, or lost, the chance of success each time is $\frac{1}{3}$. For two games the chance of success is $\frac{1}{3}\times\frac{1}{3}=\frac{1}{9}$, and for six games the chance is $(\frac{1}{3})^6$.

No matter how many events are involved the total of all the possibilities must still equal unity.

If we toss two coins they can fall in four ways—HH, HT, TH, TT—the probability of each of which is $\frac{1}{4}$. This may be seen by listing all possible ways, each of the four of which is equally probable, or by calculating the probability of each, which is $\frac{1}{2}\times\frac{1}{2}=\frac{1}{4}$. If we are not interested in the order in which the coins fall, then HT and TH are equivalent, and the addition rule gives their combined probability as $\frac{1}{2}$.

In the case of dependent events the multiplication law still applies, as long as we adjust the probabilities. Thus, the probability of dealing two aces in succession from a full pack is $\frac{4}{52}\times\frac{3}{51}$. The probability of dealing an ace and a court card is $\frac{4}{52}\times\frac{12}{51}$.

## 7.3 PERMUTATIONS AND COMBINATIONS

Suppose that there is a competition which involves picking 5 film stars from a list of 12 and putting them in order of suitability for a part. What are the chances of a random selection agreeing with that of the judges?

Here we have 5 vacancies, the first of which may be filled in 12 ways. Having chosen the first we have 11 candidates for the second place, then 10 for the third and so on. The number of ways of filling the 5 places is thus 12.11.10.9.8=95,040. This is the number of *permutations* of 12 items taken 5 at a time and is often shown as ${}_{12}P_5$.

The process of multiplying 12 by 11 by 10 etc. *down to 1* is called *factorial* 12 and is denoted by 12! Factorial n is thus n! The general formula for permutations can be expressed in this notation as ${}_nP_r = n(n-1)(n-2)\ldots(n-r+1)=\dfrac{n!}{(n-r)!}$.

In the case of permutations account is taken of the order in which the items are to appear. If we are only concerned with selecting 5 items from 12, without regard to order, we have the number of *combinations* of 12 things taken 5 at a time, shown as ${}_{12}C_5$ or $\binom{12}{5}$. The number of permutations among the 5 chosen items is 5.4.3.2.1=120, so that the number of ways in which 5 items may be chosen from 12 without regard to order is $\frac{95,040}{120}=792$. This amounts to saying that, of the 95,040 permutations, there will be 792 groups containing different items, and in each of these groups the 5 chosen items can be arranged in 120 different ways which do not concern us.

The general formula for combinations is ${}_nC_r=\frac{{}_nP_r}{r!}=\frac{n!}{r!(n-r)!}$.

Cards, dice and competitions may, at first glance, appear to be far removed from most practical problems. In fact there is often a close connection. If, for example, we wish to control the quality of the output of a factory we are interested in the chance of defective items appearing. The simple basis so far explained can be developed to provide methods essential to the analysis of statistical data.

## 7.4 THE BINOMIAL DISTRIBUTION

The formula for combinations may be combined with the multiplication law where there are a number of events and the order in which they occur is not important. We have already considered the simplest case of this (section 7.2) in connection with two coins, the probabilities of which were shown to be

| H | T | Probability |
|---|---|---|
| 2 | 0 | $\frac{1}{4}$ |
| 1 | 1 | $\frac{2}{4}$ |
| 0 | 2 | $\frac{1}{4}$ |
| | | 1 |

There is only one way in which two heads can occur and the probability is $\frac{1}{2}\times\frac{1}{2}=\frac{1}{4}$. The same applies to two tails.

There are two ways in which head and tail can occur so the total probability is $2\times\frac{1}{2}\times\frac{1}{2}=\frac{2}{4}=\frac{1}{2}$. Another way of looking at this is to regard it as the probability of any combination of events that constitutes success $(\frac{1}{2}\times\frac{1}{2})$, multiplied by the number of such combinations (2).

Probably the most useful way of tackling such problems is to treat them as a binomial expansion, where p and q are the probabilities of success and failure, totalling 1, n is the number of events being considered, and each term of the expansion gives, in succession, the probabilities of n successes, n−1 successes, n−2 successes, etc.

We know (see chapter 21) that the expansion of $(p+q)^n$ gives $p^n+np^{n-1}q+\frac{n(n-1)}{2!}p^{n-2}q^2+\frac{n(n-1)(n-2)}{3!}p^{n-3}q^3+\ldots+q^n$. It will be seen that the first term ($p^n$) is the probability of their being n successes. The second term is the probability of n−1 successes and 1 failure, $p^{n-1}q$ giving the probability of any one of such combinations of which there will be n, since the failure can occupy any of the n places. The third term gives the probability of n−2 successes and 2 failures, $p^{n-2}q^2$ giving the probability of any one of such combinations of which there will be $\frac{n(n-1)}{2!}$, since this is the number of ways in which 2 can be chosen from n, without regard to order.

In the simple case of the two coins $p=q=\frac{1}{2}$ and n, the number of events, is 2, so we have the expansion of $(\frac{1}{2}+\frac{1}{2})^2=\frac{1}{2}^2+2.\frac{1}{2}.\frac{1}{2}+\frac{1}{2}^2$, giving the terms $\frac{1}{4}$, $\frac{1}{2}$, and $\frac{1}{4}$, which indicate, in order, the probabilities of 2, 1, and 0 heads occurring.

If 6 coins were tossed the probabilities would be the terms of the expansion of $(\frac{1}{2}+\frac{1}{2})^6=(\frac{1}{2})^6+6.(\frac{1}{2})^5.\frac{1}{2}+\frac{6.5}{1.2}.(\frac{1}{2})^4.(\frac{1}{2})^2+\frac{6.5.4}{1.2.3}.(\frac{1}{2})^3.(\frac{1}{2})^3+\frac{6.5.4.3}{1.2.3.4}.(\frac{1}{2})^2.(\frac{1}{2})^4+\frac{6.5.4.3.2}{1.2.3.4.5}.\frac{1}{2}.(\frac{1}{2})^5+\frac{6.5.4.3.2.1}{1.2.3.4.5.6}.\frac{1}{2}^6$. The first term gives the probability of all six coins being heads, the second term gives the probability of there being 5 heads and one tail, etc. The probabilities work out to 0·016, 0·094, 0·234, 0·313, 0·234, 0·094, 0·016, approximately, totalling 1, as is to be expected.

Where p and q are equal, as in the case of a fair coin, the binomial distribution is symmetrical. Low probabilities of the same values occur at the start and finish, with higher probabilities in the middle. If is obvious that in the case of a coin we may substitute H for T, and vice versa, without affecting the mathematics.

In cases where p and q differ the distribution is no longer symmetrical.

Consider a 6-sided die. If this is shaken four times the probabilities of 4 sixes, 3 sixes, etc. are given by the expansion of $\left(\frac{1}{6}+\frac{5}{6}\right)^4=\left(\frac{1}{6}\right)^4+4\left(\frac{1}{6}\right)^3\left(\frac{5}{6}\right)+\frac{4.3}{1.2}\left(\frac{1}{6}\right)^2\left(\frac{5}{6}\right)^2+\frac{4.3.2}{1.2.3}\left(\frac{1}{6}\right)\left(\frac{5}{6}\right)^3+\frac{4.3.2.1}{1.2.3.4}\left(\frac{5}{6}\right)^4$.

Clearly the first item in this series is much smaller than the last, reflecting the fact that on each throw the chance of a number other than 6 is 5 times that of 6.

For any term in the series the power to which the first, or *p* value, is raised designates the number of successes indicated by that term. The power to which the second, or *q* value, is raised designates the number of failures indicated by that term.

Cases where p and q are equal lead up to the symmetrical Normal distribution (section 8.1). Where they differ greatly the series develops into the Poisson distribution (section 8.3). Both are enormously important in statistical method.

It is obvious that the mean of the binomial distribution is the number of items considered multiplied by the probability of success, i.e. np. The standard deviation can be shown to be $\sqrt{npq}$.

## 7.5.1 WORKED EXAMPLE

*Question*

A business man is trying to decide whether to promote a new product. The expected return from such promotion is vital to his decision. He thinks that, in the first year, there is a fifty-fifty chance of making a profit of £10,000.

If certain circumstances occur he may break even or may make a loss of £4,000, and he rates the chances of these equally. What is his profit expectation for the first year?

*Answer*

The probability of a profit of £10,000 is 0·5. The chance of zero profit (break even) or of a loss is thus 1—0·5=0·5, divided equally between the two possibilities, i.e. 0·25 for each.

The expected profit is thus £4,000 calculated as follows:

| | | |
|---|---|---|
| 0·5 × | 10,000= | 5,000 |
| 0·25× | 0= | 0 |
| 0·25× | —4,000= | —1,000 |
| 1·00 | | 4,000 |

[In cases where no information can be obtained about the probabilities of different happenings a version of *Bayes*' rule is generally used and we assume that all are equally probable].

### 7.5.2 WORKED EXAMPLE

*Question*

It has been estimated that 1 in 5 members of the population read a certain periodical. If members of the public are interviewed at random calculate the probabilities of the following:

*a.* That from a sample of 5 people none will be readers.
*b.* That out of a sample of 10 exactly the expectation will occur.
*c.* That from a sample of 5 there will be up to 3 non-readers.

*Answer*

Let $p=0{\cdot}2$ be the probability of readership and $q=0{\cdot}8$ that of non-readership.

*a.* The required probability is the last term of the expansion of $(p+q)^5$, i.e. $q^5=(0{\cdot}8)^5=0{\cdot}328$.

*b.* The expectation is $n\times p=10\times 0{\cdot}2=2$, and the probability of there being exactly 2 readers is given by the term of the expansion containing p raised to the 2nd power, i.e. $\frac{10.9}{1.2}p^2q^8=45\times(0{\cdot}2)^2\times(0{\cdot}8)^8=0{\cdot}302$.

*c.* We require the first 4 terms of the expansion of $(p+q)^5$, that is where $q^0$, $q^1$, $q^2$ and $q^3$ occur, i.e. $(0{\cdot}2)^5+5(0{\cdot}2)^4(0{\cdot}8)+\frac{5.4}{1.2}(0{\cdot}2)^3(0{\cdot}8)^2+\frac{5.4.3}{1.2.3}$ $(0{\cdot}2)^2(0{\cdot}8)^3=0{\cdot}2627$.

[This may also be regarded as the probability of there being 2 or more readers, which is the same as (1—probability of up to 1 reader)=1— $\frac{5.4.3.2}{1.2.3.4}(0{\cdot}2)(0{\cdot}8)^4-(0{\cdot}8)^5=1-0{\cdot}7373=0{\cdot}2627$. Tables may be obtained for this sort of calculation.]

## 7.6 PRACTICE QUESTIONS

1. For the output data of table 3.3, assuming there is no seasonal pattern, what are the probabilities that:
   *a.* A particular week's output will be less than 90 tons?
   *b.* An output of less than 70 tons will be followed by one of over 100 tons?
   *c.* A particular week's output will be either 70–79 or 100–109 tons?

2. For the output of machine B of table 3.4 assume that any item less than 1·003 inches is scrap.
   *a.* Out of a batch of 500 items how many scrap items would be expected?
   *b.* What is the probability of exactly 2 scrap items from 10?
   *c.* What is the probability of up to 2 scrap items from 5?

3. For the bonus data of table 2.6 calculate the following probabilities:
   *a.* That a person selected at random receives less than £5 bonus.
   *b.* That two persons (random) will both receive £5–£6 bonus.
   *c.* That any random group of 5 people will contain just 3 persons with a bonus of £6–£7.

   State any assumptions implied in your working.

4. A firm wishes to devise a simple publicity competition, with a cash prize offered for a person selecting a number of sayings from a list and placing them in correct order. Advise on the statistical implications of this and suggest means of avoiding pitfalls.

# 8

# Some Important Distributions

## 8.1 THE NORMAL DISTRIBUTION

The *normal distribution*, which is also referred to as the *normal curve of error* and the *Gaussian curve*, is probably the most important distribution in statistics. It is the well-known symmetrical, bell-shaped curve shown in diagram 8.1, and applies to many natural phenomena, where the bulk of

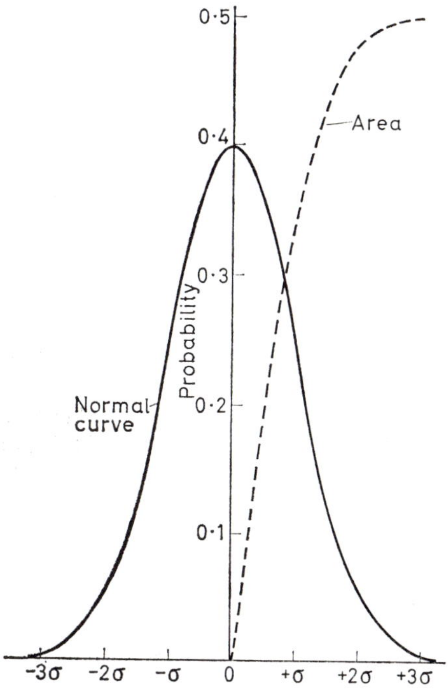


Diagram 8.1. *Normal distribution*, with curve showing area from mean to any given value

the items cluster round a central value and their number falls off rapidly at the more extreme values.

We have seen in the previous chapter that the successive terms of the binomial distribution show the probabilities of n, n−1, n−2 etc. successes, and that each term is made up from $\frac{n!}{r!(n-r)!}p^r q^{n-r}$. This is sometimes referred to as *Bernoulli's Law*.

When p=q this distribution is symmetrical and from it can be developed the continuous function for the normal distribution, the formula for which is

$y = \frac{1}{\sqrt{2\pi}} e^{-\frac{1}{2}z^2}$, where z is the original data expressed as deviations from its mean and divided by its standard deviation (see section 6.5.3). The meaning of $e$ is explained at 21.3.4; it is a constant and equals 2·71828.

The total area under the normal curve is one, and the area between any two values of z indicates the probability that these values, or anything between them, will arise.

Diagram 8.1 shows the shape of the normal curve. It also indicates the area from its mid-point (or average) to any selected point, by the broken line. Since the curve is symmetrical half the area lies on each side, so the area from the extreme left to a point on the right can be found by adding 0·5 to these values. The *tails* of the distribution can be found by subtracting these values from 0·5.

Students should make themselves familiar with tables of the area under the normal curve, which sometimes show the area from the mid-point, as above, sometimes from the extreme lower end (totalling one finally), and sometimes show the tails of the distribution.

It is useful to remember that approximately 95% of all values lie within a range of $2\sigma$ on each side of the mean (accurately $1{\cdot}96\sigma$ gives this percentage). This is often called the *5% confidence level,* since we know that only 5% of the items will occur outside this range by chance.

A range of $3\sigma$ on each side of the mean accounts for 99·74% of all the items, and this is used as a very rigid indication of whether or not an item comes from the population.

A less exacting level is at 1%, which includes 99% of all the items and lies at $2{\cdot}58\sigma$ on each side of the mean.

The values just given refer to the cut-off of both tails of the distribution. If we are concerned with the upper or lower tail only the 5% level is at $1{\cdot}65\sigma$ and the 1% at $2{\cdot}33\sigma$.

We saw (section 6.5.3) that data can be standardised so that, if the distribution is normal, the same set of tables can be used for an infinite variety of data. This section and table 6.7 should be referred to if the reader has any doubts about what is involved. See also summary tables 1 and 2.

From the output data of table 6.6 we have found the mean to be 91·4 and the standard deviation to be 10·7. We may deduce a number of things from these facts.

Is it likely that an output of 125 would arise? The answer is found by calculating the difference between 125 and the mean ($125-91{\cdot}4=33{\cdot}6$) and expressing this in terms of the standard deviation $\left(\frac{33{\cdot}6}{10{\cdot}7}=3{\cdot}1\right)$. Since deviations of $3\sigma$ or more from the mean occur less than 3 times per 1,000 we conclude that it is very unlikely that such a value could occur. This sort of calculation, in conjunction with sampling, is the basis of one of the methods of quality control outlined in chapter 10.

What percentage of outputs will lie between 80 and 100? In standard terms 80 is $\frac{80-91{\cdot}4}{10{\cdot}7}=-1{\cdot}065$, while 100 is $\frac{100-91{\cdot}4}{10{\cdot}7}=0{\cdot}804$. These lie on each side of the mean. Area tables of the normal curve show that the area from the mean to $-1{\cdot}065$ (the same as to $+1{\cdot}065$) is 0·3577, and that from the mean to 0·804 is 0·2881. The sum of these gives the required probability, 0·6458. Thus, about 65% of outputs would occur within the stated limits.

What percentage of output will lie between 100 and 110? We have done the standardising calculation for 100; that for 110 gives $\frac{110-91\cdot4}{10\cdot7}=1\cdot74$. The areas from the mean to 100 and 110 respectively are 0·2881 and 0·4591. Since both lie on the same side of the mean the required probability is the difference, giving 0·171. Thus 17% of the output will occur between the stated values.

Students will find it useful to draw small sketches of the normal curve in dealing with problems such as those just illustrated and to mark the values concerned with vertical lines.

The normal curve probabilities are frequently used in assessing the results of experiments and investigations. This is taken further in the chapter on sampling (chap. 9) and the chapter on more advanced aspects (chap. 20). Briefly, we are concerned with whether a result could reasonably have occurred by chance or whether it is so unlikely that it must be regarded as *significant*. The probability level to be considered as significant, say 5% or 1%, is left to the investigator.

It is possible to make mistakes in these judgments. A *type 1 error* occurs when we wrongly reject a hypothesis because the result seems explicable in terms of chance; a *type 2 error* implies acceptance of an incorrect hypothesis because the results seem to indicate that it is too exceptional to be explained as a random occurrence.

## 8.2 THE POISSON DISTRIBUTION

This approximation to the binomial distribution applies when the probability of an event is small. The probabilities p and q in this case are very different and the result is a highly skewed distribution.

The number of items (which must be large) is denoted by *n*, and the expected number of successes by $m=np$. This is the average number of successes to be expected in a large number of trials. The Poisson distribution is then

$$e^{-m}(1+\frac{m}{1!}+\frac{m^2}{2!}+\frac{m^3}{3!}+\frac{m^4}{4!}+\ .\ .\ .\ .)$$

The first term gives the probability of no successes (we may regard 1 as being $m^0$), the second, where the power of m is 1, gives the probability of 1 success etc., the power of m indicating the number of successes to which that term of the expansion relates.

Diagram 8.2 shows the Poisson distribution for various values of m. It should be noted that it is not really a continuous distribution. It exists only for the whole numbers on the horizontal axis. We cannot have a fraction of an event occurring. Summary table 3 gives values of $e^{-m}$.

Suppose that a production process is resulting in 1% scrap. We wish to calculate the probability of samples of 100 items containing different numbers of defectives. Here $n=100$ and $p=0\cdot01$, so that $m=np=1$. The probabilities are as follows:

$$e^{-1}\,(1+\frac{1}{1}+\frac{1^2}{2}+\frac{1^3}{6}+\frac{1^4}{24}\,.\ .\ .\ .\ .\ .\ .), \text{ which,}$$

since $e^{-1}=0{\cdot}3679$ (from tables), gives:

| | | |
|---|---|---|
| 0 defectives | (first term) | 0·3679 |
| 1 defective | (second term) | 0·3679 |
| 2 defectives | (third term) | 0·1840 |
| 3 defectives | (fourth term) | 0·0613 |
| 4 defectives | (fifth term) | 0·0153 |
| | total to fifth term | 0·9964. |

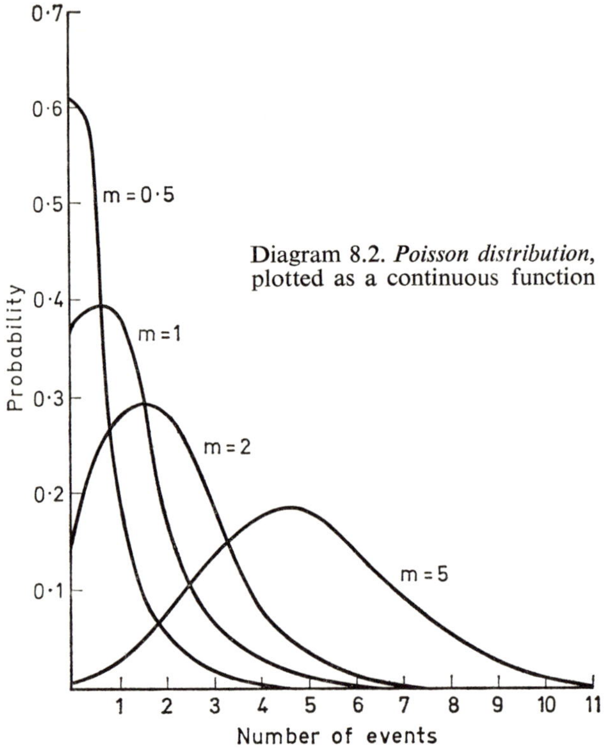


Diagram 8.2. *Poisson distribution*, plotted as a continuous function

The Poisson distribution rapidly tails off, its terms soon becoming insignificant. It is the basis of *acceptance sampling* (see chapter 10). It is also of use in *stock control* when only a few units are sold per unit of time, and in *queue theory*, where a number of small intervals of time is considered, $p$ being the chance that an *arrival* will occur, and $m$ being the number of arrivals on average over the number of time units considered.

If observations show that, on average, there are 5 arrivals each hour, and we imagine this length of time split into a large number of units of time, then m=5 and the probabilities of 0, 1, 2 etc. units arriving per hour are the terms of the expansion $e^{-5}(1+5+\frac{5^2}{2!}+\frac{5^3}{3!}+\frac{5^4}{4!}+\ .\ .\ .)$, i.e. 0·0067, 0·0337, 0·0842, 0·1404, 0·1755, 0·1755, 0·1492, 0·1066 etc. Once we have the first term in the series, obtained from tables of the value of $e^{-m}$, the others are derived from it by multiplying the previous result by 5, $\frac{5}{2}$, $\frac{5}{3}$, $\frac{5}{4}$ etc., successively.

It has been seen that the mean of the Poisson distribution is the value *m* which is used to distinguish the various curves. The standard deviation can be shown to be $\sqrt{m}$.

## 8.3 THE BETA DISTRIBUTION

This will not be considered in detail. In shape it has affinities with the normal distribution.

The chief interest is its use in connection with scheduling *networks* of the *PERT* (Project Evaluation and Review Technique) type. These show how a project is built up to completion from a number of *activities* (e.g. laying foundations, building the walls, glazing, roofing etc. in the case of a building project) shown as arrows linking the various stages of completion, which are known as *events* and are shown as circles. Estimation of the times required to complete the activities is vital to this analysis.

Three times may be attached to each activity:

1. $t_s$ which is the shortest time if all goes well,
2. $t_m$ which is the most likely average time, and
3. $t_l$ which is the longest time if difficulties crop up, except major ones which would kill the project.

There are reasons for supposing the beta distribution gives a satisfactory description of this sort of situation, and the statistically expected time (T) and its standard deviation (s) are easily found from the formulae

$$T=\frac{t_s+4t_m+t_l}{6}$$

and $$s=\frac{t_l-t_s}{6}$$

T is the value used in doing the network calculations, and the values of s enable the probability of completion by a given date to be estimated.

Table 8.1 shows data of this type with a very limited, and unrealistic, number of activities. This PERT or critical path network is referred to as a practical example in chapter 16, where the network drawing is shown.

Table 8.1. *Statistical treatment of PERT network*

(See chapter 16 for descriptive treatment and diagram)

| Activity | $t_s$ | $t_m$ | $t_l$ | T | s | $s^2$ |
|---|---|---|---|---|---|---|
| *1—2 | 7 | 14 | 21 | 14 | 2·33 | 5·43 |
| *2—3 | 2 | 3 | 10 | 4 | 1·33 | 1·77 |
| 3—4 | 2 | 3 | 4 | 3 | 0·33 | 0·11 |
| *3—5 | 2 | 2 | 8 | 3 | 1·00 | 1·00 |
| 4—6 | 3 | 4 | 11 | 5 | 1·33 | 1·77 |
| 6—7 | 5 | 7 | 15 | 8 | 1·67 | 2·79 |
| *5—7 | 20 | 28 | 42 | 29 | 3·67 | 13·47 |
| *7—8 | 4 | 7 | 10 | 7 | 1·00 | 1·00 |

The longest path through the network, called the *critical path*, connects activities 1—2—3—5—7—8, which are starred. Addition of these starred items (T) gives a statistically expected total time of 57 days required for completion.

The square of items in the standard deviation column (s) gives the variances ($s^2$), which have been added for the critical path, giving 22·67. The standard deviation for the whole path is thus $\sqrt{22{\cdot}67}=4{\cdot}76$ days.

We can now use the tables of the normal curve to find the probability of completion within any chosen number of days beyond the estimated total time.

If we wish to know the probability of completion within, say, 5 days then the standard normal variable (z) is $\frac{5}{4{\cdot}76}=1{\cdot}05$, for which the single-tail probability is about 85%.

For completion within 1 day of the estimated time $z=\frac{1}{4{\cdot}76}=0{\cdot}21$, for which the single-tail probability is about 58%.

If we do similar calculations for any selected time for all paths through the network and multiply their probabilities we obtain an estimate of the probability of completion of the whole project in that time. This is rather dubious since many paths have activities in common and, with a small network such as this, the sample is small for the distribution to be regarded as normal.

## 8.4 SIMULATION

*Simulation* is an important tool of *operational* research. This latter studies business and management situations, attempting to build a mathematical description or *model*, which can then be tested to see the effects of events which are likely to arise in practice.

An important feature of many of these studies is the introduction of chance or random elements, instead of using rigidly predetermined events. This makes the studies much more realistic.

*Top hat* and *Monte Carlo* techniques enable this to be done and imply that the data is drawn from a hat or selected by some other random method such as a table of random numbers (table 9.1).

We may find the relative frequency of occurence of various events by listing these and keeping a frequency tally, or by activity sampling (section 9.2). On the other hand we may assume that the events follow the pattern of, say, the normal or Poisson distributions, using the tables to give us the relative frequencies.

Simulation has the important characteristic, shared with real life, that we never know precisely what is going to happen at different stages of the simulation, though the long-term average outcome may sometimes be obtained from a formula in simpler cases.

*Queues* afford good examples of simulation. The characteristic of the queue situation is a pattern of *arrivals* and of *service* times which may be connected by a *queue*.

If there are, say, 3 arrivals on average each hour then the average number (m) arriving in any 5-minutes is $3\times\frac{5}{60}=0{\cdot}25$. If the Poisson distribution applies the probabilities of 0, 1, 2 etc. arriving in any 5 minute interval will be the terms of $e^{-0{\cdot}25}\left(1+0{\cdot}25+\frac{(0{\cdot}25)^2}{2!}+\frac{(0{\cdot}25)^3}{3!}+\ldots\right)$. These are listed below and, alongside, are shown the 2-figure random numbers (from 1 to 99) which could be used to simulate the arrival pattern.

| *No. Arriving* | *Probability* | *Random Nos.* |
|---|---|---|
| 0 | 0·78 | 1—78 |
| 1 | 0·19 | 79—97 |
| 2 | 0·02 | 98—99 |
| 3 | 0·00 | — |

We can have a sheet of paper, each line representing 5 minutes. As numbers are drawn from a hat, or better read from a random number table, we record, for each 5 minutes, how many arrivals occur. Letters can be used to distinguish between these.

A study of service times might have revealed the length-of-service pattern to which random numbers would be similarly allocated, so that we can determine how long each simulated service will take. If service facilities are free an arrival goes directly into service, otherwise it joins the queue, the length of which can be noted line by line.

Try your hand at drafting a blank table suitable for recording such a simulation.

Such manual simulation is very slow and digital computers can give greater speed and enable very complex situations to be simulated. A random number table can be read into the computer memory but it is easier to generate *pseudo-random* numbers as part of the computer program.

### 8.5.1 WORKED EXAMPLE

*Question*

The daily sales of a particular article are observed to be as follows over 70 days. What level of stock should ensure that the business does not run out of stock more than 5% of the time?

| | | Calculation | | |
|---|---|---|---|---|
| No. of Items Sold | Frequency (f) | Dev. (D) from $34\frac{1}{2}$ in 10s | fD | $fD^2$ |
| 10—19 | 4 | −2 | −8 | 16 |
| 20—29 | 13 | −1 | −13 | 13 |
| 30—39 | 42 | 0 | 0 | 0 |
| 40—49 | 10 | 1 | 10 | 10 |
| 50—59 | 1 | 2 | 2 | 4 |
| | 70 | | −9 | 43 |

*Answer*

The last three columns show the calculation of the mean and the standard deviation.

$$\bar{x}=34\tfrac{1}{2}-\frac{9}{70}\times 10=33.2 \text{ items}$$

$$s=\left(\sqrt{\frac{43}{70}-\left(\frac{9}{10}\right)^2}\right)\times 10=7{\cdot}8 \text{ items.}$$

The 5% level, cutting off only one tail of the normal distribution, is at 1·65s above the mean, i.e. at $33 \cdot 2 + 1 \cdot 65 \times 7 \cdot 8 = 46$.

If 46 items are in stock at the start of any day demand should be met 95% of the time, assuming that the pattern of demand remains as stated.

### 8.5.2 WORKED EXAMPLE

*Question*

A furniture shop sells 3 units of a particular item each week. The policy is to save storage space by keeping stocks low and only 3 items are in stock at the start of each week, none other being available during the week. What is the probability of being unable to meet consumer demand?

*Answer*

We can treat this as an application of the Poisson distribution where m=3 (the amount demanded).

The probability of up to 3 items being demanded is given by

$$e^{-3}\left(1+\frac{3}{1!}+\frac{3^2}{2!}+\frac{3^3}{3!}\right)=0{\cdot}04979+0{\cdot}14937+0{\cdot}22406+0{\cdot}22406=0{\cdot}64728.$$

The probability of running out of stock is thus $1-0{\cdot}65=0{\cdot}35$.

## 8.6 PRACTICE QUESTIONS

1. The mean height of a group of 400 male adults is 68·16 inches and the standard deviation is 2·54 inches. Assuming the distribution is normal estimate the number *a.* over 6 ft., *b.* less than 5 ft. 4 in., *c.* between 5 ft. 6 in. and 6 ft.

2. From the data of qu. 1 estimate *a.* the heights of the lower and upper quartiles, *b.* the minimum height of a person likely to be in the highest 10%, *c.* the maximum height of a person likely to be in the lowest third of the group.

3. It has been found on average that a clerk makes 2 errors for every 100 calculations of a certain kind which he performs. In a particular day he does 500 such calculations. Estimate *a.* the expected number of errors, *b.* the probability of there being exactly 8 errors, *c.* the probability of not more than 6 errors.

4. The average number of persons calling at an enquiry office is 6 every 10 minutes. Calculate the probability *a.* of exactly 3 persons appearing in any 10 minutes, *b.* of more than 2 persons appearing in 10 minutes, *c.* of 3 persons appearing in 5 minutes.

5. The average of a normally distributed variable is 50 and 8% of the values are less than 40. Calculate *a.* the standard deviation, *b.* the percentage of items likely to exceed 55.

6. Explain the differences between the normal and Poisson distributions, giving examples of the circumstances to which each might be appropriate in your particular field of studies.

# 9

# Sampling

## 9.1.1 THE USES OF SAMPLING

Sampling implies an attempt to assess some properties of a larger number of items by a study of similar properties of a smaller number.

The term *universe* or *population* is used to refer to the bulk from which the sample is drawn.

Sampling is useful in a number of situations:

a. Where the whole of the population is not accessible, so that sampling is the only means available.

b. Where a study of the whole of the population would be too expensive or would take too long.

c. Where the attribute being sampled would involve the destruction of the item, such as occurs in the destructive testing of manufactured items.

Our approach to life is generally based on sampling though we do not generally recognise this. From birth the child is subject to a number of experiences from which he draws a picture of the universe in which he lives. This pattern is modified as the sample is enlarged and newer experiences become available.

## 9.1.2 SAMPLING ERRORS AND STANDARD ERROR

Generally in statistics we regard most data as being a sample drawn from a wider universe. This sample is liable to *sampling variations* or *errors* arising from the fact that another sample set of data would not yield precisely the same result.

These sampling errors must be distinguished from the other errors which can occur when data is collected, such as those arising from wrong counts, incorrect classification of items and the wrong interpretation of questions.

We have noted elsewhere that the statistical convention is to denote a population parameter by a Greek letter and the sampling approximation to this by a normal letter. Thus we have $\mu$ (mu) for the population mean and $\bar{x}$ for the sample mean; $\sigma$ (sigma) for the population standard deviation and s for the sample standard deviation; $\rho$ (rho) for the population coefficient of correlation and r for the coefficient derived from a sample (see chapter 12). It is also usual to denote the number of items in the universe by N and the number in the sample by n.

The term *standard error* is often used to denote the standard deviation as between the means of the various samples which could arise.

## 9.2 SIMPLE RANDOM SAMPLING. SAMPLING FRAMES

The approach of *simple random sampling* is to devise a sample so that each item in the universe has an equal chance of being included in it. In other words care is taken to avoid any bias in the selection of items. This is often called *probability sampling*.

Elimination of bias generally involves ruling out any element of choice on the part of the person selecting the items for the sample, and also the destruction of any links in the data which may cause one item to have a greater chance of being chosen once another item has been selected.

If we wish to assess the condition of stock items in a shop or factory it would be unwise to instruct a person to take, say, 20 tins for inspection, since he might tend to choose the 20 most accessible. This might be representative or not, depending upon the procedure used for stacking the stock; with some methods they would represent only the latest items. If fraud was present they might represent those items which provide a 'front' for mostly useless stock. We might instruct the individual to take a tin from each shelf or to count up to certain specified tins, selected to give a random effect.

A table of *random numbers* can be used. This is simply a list of numbers derived in such a way that there is no connection between them. Table 9.1 shows part of such a list of two figure numbers, which have been obtained by drawing numbered discs from a box, each disc being returned after its number has been noted and the box being shaken. They are simply read in order starting at any point of the table and they indicate which items should be taken. They may be read singly as 3, 6, 5 etc. or in any combination, e.g. 36, 53, 80, or 365, 380, 137 etc. depending on the degree of precision required.

Table 9.1. *2-figure random numbers*

| | | | | | | | | | |
|---|---|---|---|---|---|---|---|---|---|
| 36 | 53 | 80 | 13 | 75 | 86 | 58 | 93 | 67 | 00 |
| 94 | 44 | 58 | 50 | 16 | 20 | 47 | 24 | 12 | 83 |
| 58 | 74 | 95 | 08 | 94 | 97 | 29 | 38 | 20 | 97 |
| 75 | 86 | 38 | 60 | 87 | 56 | 25 | 28 | 27 | 57 |
| 42 | 16 | 76 | 40 | 81 | 16 | 88 | 73 | 59 | 81 |
| 00 | 63 | 37 | 93 | 76 | 17 | 28 | 11 | 15 | 66 |
| 82 | 85 | 45 | 46 | 99 | 80 | 45 | 42 | 90 | 32 |
| 94 | 71 | 66 | 51 | 85 | 12 | 82 | 37 | 50 | 43 |
| 21 | 92 | 59 | 33 | 49 | 77 | 07 | 06 | 56 | 02 |
| 90 | 02 | 19 | 31 | 71 | 03 | 39 | 37 | 03 | 50 |

Such an approach is used in *activity sampling* where the pattern of activities of an individual or machine over a period of time is to be assessed. One way of doing this would be to have an observer recording precisely how long a man or machine is employed on different activities. The sampling approach involves observing how he is occupied at points of time and, from some hundreds of such observations, deriving the overall pattern. The random numbers can be used to determine the times at which the observations should be made.

The term *ratio delay* is often used when this technique is applied to machine running time and stoppages, which may be analysed by cause.

If we read the numbers in two's across and work commences at 9 a.m. then the first observation would be made at 9.36 a.m., the second 53 minutes after this at 10.29 a.m., the third 80 minutes after this at 11.49 a.m., the fourth 13 minutes after this at 12.02 p.m. Time intervals may be selected to get the size of sample required. If we wanted more observations we might work in half minute intervals so that the first observation would be at 9.18 a.m. etc.

We may use this technique to sample the bonus data of table 2.1. Let us read down the columns of random numbers for a change. The first single number is 3, so we take the third item in the first column of the bonus data, i.e. 4·75. We then take the sixth item after that (4·15), the ninth item after that (4·50), the fourth item after that (5·65), and so on to the end of the table. This gives 29 items sampled. If we need 30 items we can continue through the table of data again. The total of these 30 items is 162·30, giving an average of 5·41, compared with the actual average of all the numbers of 5·38. In this case a zero random number has been omitted if it occurs.

In the case of *market research*, involving consumer surveys, and *sociological research*, generally, it is much more difficult to obtain a suitable random sample than is at first appreciated. If we instruct the interviewer to obtain answers from a hundred individuals the probability of bias is very great. If she calls on the nearest houses the bias arises for a number of reasons:

a. There is bias in the area being selected.
b. Only people who are at home can possibly respond so that the sample probably biased towards housewives and older people during the day.
c. If only co-operative people are assessed this may vitiate the whole sample.

Point c. is one of particular difficulty since only Government investigations are normally compulsory. With private investigations a heavy *non-response* may be reduced by an approach which arouses the interviewee's interests. In some investigations people who are not selected as part of the sample may clamour to be included feeling that they are missing something if they are not interviewed. Clearly, in selecting people, sources of information which are biased should be avoided. Probably the best known example of this is the telephone directory which does not include a representative selection of the population despite the increased popularity of the telephone. This does not rule it out in all cases, however. It depends on the population with which we are concerned. For some investigations this might well be restricted to telephone subscribers.

For general survey work the *Register of Electors* is regarded as a basic source of information. Rating records are also useful.

The term *sampling frame* is often used for the lists of individuals, business, customers etc. from which the sample is drawn. Such lists should include all the units in the relevant population, which may be difficult to achieve.

## 9.3.1 OTHER SAMPLING TECHNIQUES

Selection of a truly random sample of a universe may be difficult. A greater accuracy at lower costs may often be obtained by some other approach which deliberately selects the items to be included according to certain characteristics which are known to exist in the universe. The sample becomes a microcosm of the universe in these respects, and we assume that it will behave similarly with regard to the unknown characteristic. This approach is often combined with a random element. The best known alternatives to simple random sampling are:

a. stratified and quota sampling.
b. cluster sampling and multi-stage sampling.
c. systematic sampling.

### 9.3.2 STRATIFIED AND QUOTA SAMPLING

Here the population is split into groups or strata. As an example we may work on age groups or income groups or on areas of residence. From each of these groups a sample can be obtained on a random basis, the number being *proportional* to the relative importance of that group in the population. With *cross-stratification* a number of characteristics are handled.

The importance may depend on the frequency with which the groups occur. Often, however, a small number of units account for the bulk of the cases with which we are concerned. A few large firms employ more workers than do a much greater number of small firms (see table 3.11). In cases such as this we may take all the large firms, a proportion only of the not-so-large, and so on, the proportion declining with smaller firms, ending, perhaps, with none of the tiny firms in our sample. This could be used for data on labour conditions. This is generally termed using *variable sampling fractions.*

In *optimum allocation sampling* a larger proportion of items is taken from strata of particular interest or where an issue is finely balanced.

*Quota sampling* is a modification of stratified sampling where the proportions of the different strata are pre-determined but the choice of the actual items contributing to these strata is left to the interviewers. They have a quota of interviews to make but do not have to face the time-consuming problem of previously selected persons not being available and thus requiring several calls to avoid non-response.

### 9.3.3 CLUSTER AND MULTI-STAGE SAMPLING

The approach is to split the population into groups and then use sampling to select the group which will be used.

Let us take the bonus data. We can regard each column of 25 items as a cluster and use a 6-sided die to decide which column to take. If it comes up 4, then we take the fourth column of which the average is 5·41. This uses all the 25 items in the cluster. Often, having determined the cluster, we should divide it into smaller clusters from which samples would be taken or we could sample from the individual items in it. This gives rise to the term *two-stage sampling* sometimes called *sub-sampling.*

This idea can be extended to several stages of sampling, called *multi-stage sampling.* We may select areas of the country, then towns within the areas, then streets in the towns, then houses in the streets etc., all by means of sampling.

### 9.3.4 SYSTEMATIC SAMPLING

This starts from a randomly selected item and then takes every nth item, n being chosen to give the number required in the sample. With the 150 bonus items a sample of 20 would require every seventh item to be selected. If we work from the third item (randomly chosen) we have 4·75, 5·80, 4·50, 3·35, etc., and the average of these twenty items is 5·48. A danger with systematic sampling is that the items chosen may represent some underlying pattern in the data, e.g. every tenth house in a terraced district might be an end house.

### 9.4.1 SIGNIFICANCE AND SAMPLE SIZE. CENTRAL LIMIT THEOREM

Here we are dealing with an extension of the ideas developed in chapter 8. There we were concerned with individual items from a normal population and the probabilities of their differing from the population mean by stated amounts. Our problem now relates to the means of samples drawn from such a population. What are the probabilities of such means differing from the population mean by stated amounts?

From a common sense point of view we should consider that such sample means, since they result from the averaging of individual items, will be less widely dispersed than the individual items.

The rest of this chapter is concerned with developing the simple mathematics of this concept.

It can be shown that the means of samples of n items drawn from a normal population will be distributed about the mean of the population with a standard deviation of $\frac{\sigma}{\sqrt{n}}$, where $\sigma$ is the standard deviation of the population. This is known as the *central limit theorem.*

It is usual to divide consideration of sampling into cases where a *variable* is being considered (for example heights of people or the weight of a manufactured product) or where we are dealing with an *attribute* (for example the proportions of people who like or dislike a product or the quantities of manufactured output which are good or are reject).

### 9.4.2 VARIABLES

Let us refer once again to table 2.1. We may regard the 150 items in it as a population. Each of the six columns of 25 items may be regarded as a cluster for sampling purposes. We have previously calculated the mean of the population to be 5·38 and the standard deviation as 0·8. With samples of 25 this implies that the standard deviation of the distribution of the sample means ($\sigma_{\bar{x}}$) is $\frac{\sigma}{\sqrt{n}}=0{\cdot}16$. This may be referred to as the *standard error* (*s.e.*)

The properties of the normal curve are such that 95% of sample means should lie in the range $5{\cdot}38 \pm 2 \times$ s.e., i.e. from 5·09 to 5·73.

Calculation of the means of the six columns yields values of 4·90, 5·51, 5·48, 5·41, 5·36 and 5·64, from left to right. All except the first occur within the range just calculated. Division into the six columns represents only one of many random ways of obtaining samples from the data. It is a useful exercise for the reader to estimate how many random combinations of 25 items are possible. Use the techniques of 7·3 and take account of two different approaches. We may assume either that once an item has been drawn it is no longer available or, alternatively, that each item is 'replaced' after it has appeared in a sample, so that it can be used again.

When we take a sample we are often concerned with whether the mean of this sample could reasonably have been obtained from a population having a stated mean. In the case above five of the six columns seem to be in the range expected, at the 5% level of significance, given that the population mean is 5·38. It seems that the first column differs significantly from this population.

Suppose we wish to test whether the sample mean of 5·41 could reasonably have arisen from a population having a mean of 5·7 and a standard deviation of 0·8. At the 5% confidence level 95% of the sample means should lie in the range $5{\cdot}7 \pm 2 \times \sigma_{\bar{x}} = 5{\cdot}7 \pm 2\frac{\sigma}{\sqrt{n}} = 5{\cdot}7 \pm 0{\cdot}32$, i.e. from 5·38 to 6·02. The value of 5·41 is within this range so the assumption is not unreasonable. If the assumed population mean had been 5·8 the range would have been from 5·48 to 6·12 and we should have to say that the sample value of 5·41 differs significantly from this.

An important application of this occurs with experimental results. Is the average yield of a crop after using certain fertilisers significantly different from the average obtained without them? Do the average marks obtained by the pupils of a particular school differ significantly from those of all children sitting the examination?

We may occasionally be in error in our judgment. The *type 1 and type 2* errors mentioned in chapter 8 may occur. In quality control (section 10.4) the terms *producer's* and *consumer's risks* are used when such errors occur.

The larger the sample the smaller will be the standard error. This is equivalent to saying that the *precision* of the sample increases according to the square root of the number of items included. Since the cost of sampling may well increase in direct proportion to the number of items in the sample a point will be reached with the *cost* of extra precision will be too great.

It will be seen that if n is very large, approaching infinity, the standard error becomes negligible. This merely implies that our samples include the whole population, so we should not expect any variation. A disadvantage of very large samples, apart from the cost, is that the volume of work and the fact that the collection of data may involve thousands of people can lower the standard of accuracy of the investigation.

The size of sample (n) for any predetermined degree of precision may be readily calculated. We decide what difference may have to be tested and equate this to $\frac{\sigma}{\sqrt{n}}$ multiplied by the factor appropriate to the chosen level of significance. For the 5% level this factor will be 2 (or 1·96 to be exact); at the 1% level it will be 2·58 (2-tail tests).

Suppose the population mean is 85 units and the standard deviation is 2·5 units. We wish to estimate the smallest sample which will indicate the significance (at the 5% level) of a sample mean differing from the population mean by 0·1 units.

We have $2 \times \frac{2{\cdot}5}{\sqrt{n}} = 0{\cdot}1$ from which n=2,500.

If a sample of 2,500 items revealed an average outside the range $85 \pm 0{\cdot}1$, i.e. from 84·9 to 85·1 we should regard it as significantly different.

The significance of larger changes would require a smaller sample. If the possible change were 1·0 we should have $2\times\frac{2\cdot5}{\sqrt{n}}=1\cdot0$, from which n=25, which is one-hundredth the size of the previous sample.

### 9.4.3 ATTRIBUTES

In this case we are concerned with the number or proportion of items displaying a particular property. This is similar to the concept of success or failure in the language of probability. The approach is very similar to that just used for variables.

If p and q designate the proportions of success and failure then the numbers of successes and failures will be distributed with a standard error of $\sqrt{npq}$ when we take samples. If we are dealing with proportions instead of actual numbers then the standard error is $\sqrt{\frac{pq}{n}}$.

Suppose a survey of 400 people shows that 160 like a product and 240 dislike it. Is this evidence of the unpopularity of the product? If the population is evenly balanced with regard to opinion of the product there would be 200 people in each category and the probability of like or dislike (equivalent to success or failure) would be p=q=0·5.

The standard error of sampling would be $\sqrt{400\times0\cdot5\times0\cdot5}=10$. At the 5% level of confidence we might expect samples to occur in the range $200\pm2\times10$, or from 180 to 220. Our sample is thus significantly different from an evenly balanced population.

If we work in terms of proportions then the standard error becomes $\sqrt{\frac{0\cdot5\times0\cdot5}{400}}=0\cdot025$, so that samples might be expected from $0\cdot5-2\times0\cdot025$ to $0\cdot5+2\times0\cdot025$, i.e. from 0·45 to 0·55. The sample proportions are $\frac{160}{400}$ and $\frac{240}{400}$, i.e. 0·4 and 0·6, so this approach yields precisely the same conclusion.

The size of sample needed to descriminate in a particular case can be determined in the same way as for variables.

Suppose the population values of p and q are assumed to be 0·8 and 0·2 respectively. For a sample of size n the standard error would be $\sqrt{\frac{0\cdot8\times0\cdot2}{n}}=\frac{0\cdot4}{\sqrt{n}}$. If we wish to pick out significant differences of 0·01 in the values of p and q (at the 5% level) then $2\times\frac{0\cdot4}{\sqrt{n}}=0\cdot01$, giving n=6,400. A sample of 6,400 items should give the precision required.

### 9.4.4 SAMPLING ESTIMATES

So far our chief concern has been the testing of hypotheses by means of a sample. We have assumed a population mean and a value for the standard deviation.

The aim of sampling is often to estimate these parameters for the population.

With samples of about 30 or more items we can take the sample value and indicate the range on each side which would cover other possible samples. The population mean should lie in this range.

If a survey of 100 items showed values of p and q of 0·4 and 0·6 the standard error would be $\sqrt{\frac{0{\cdot}4\times0{\cdot}6}{100}}=0{\cdot}049$. The value of p for the population, at the 5% level of significance, will be $0{\cdot}4\pm2\times0{\cdot}049$ or from 0·302 to 0·498.

Similar considerations apply in the case of sampling of variables.

If the 150 bonus items of table 2.1 are considered as a sample from a larger set of items then, in default of any further information as to the population standard deviation, we must use the value obtained from the sample (0·8). The standard error is thus $\frac{0{\cdot}8}{\sqrt{150}}=0{\cdot}065$. The mean of this sample is 5·38. At the 5% significance level we may say that the population mean lies between $5{\cdot}38-2\times0{\cdot}065$ and $5{\cdot}38+2\times0{\cdot}065$ or from 5·25 to 5·51.

## 9.4.5 FINITE POPULATIONS

Up to now we have assumed that sampling is done from an infinite population or one so large as to approximate to this, or that it is done with replacement, so that the proportions in the population are not disturbed by the sampling process.

In cases where the population is finite and where once an item is drawn for the sample it is lost to the universe (thus altering its properties) we should adjust the estimate of the standard error to $\frac{\sigma}{\sqrt{n}}\sqrt{\frac{N-n}{N-1}}$, where N is the size of the universe and n is the number in the sample.

In the case of the cluster of 25 items drawn from the bonus data this would alter the estimate of the standard error to $\frac{0{\cdot}8}{\sqrt{25}}\sqrt{\frac{150-25}{150-1}}=0{\cdot}15$. We should expect the mean of the population (using the fourth column as our cluster) to lie between $5{\cdot}41-2\times0{\cdot}15$ and $5{\cdot}41+2\times0{\cdot}15$, i.e. from 5·11 to 5·71.

## 9.4.6 TWO SAMPLES

A situation of some practical importance occurs when there are two samples and we wish to test if they can be regarded as drawn from the same population.

In this case, if the samples are of size $n_1$ and $n_2$, with means $\bar{x}_1$ and $\bar{x}_2$, and standard deviations $s_1$ and $s_2$, it can be shown that the standard error of the difference between the means $=\sqrt{\frac{s^2_1}{n_1}+\frac{s^2_2}{n_2}}$.

Suppose samples of 100 and 400 items reveal means of 20·9 and 22·0 and standard deviations of 4·2 and 4·4 respectively. The difference between the means is $22{\cdot}0-20{\cdot}9=1{\cdot}1$. The standard error of this difference is $\sqrt{\frac{4{\cdot}2^2}{100}+\frac{4{\cdot}4^2}{400}}=0{\cdot}47$. The difference is thus $\frac{1{\cdot}1}{0{\cdot}47}=2{\cdot}3$ times the standard error, so it is significant at the 5% level. Tables of the area under the normal curve show that such a value would occur only about 2% of the time by chance.

Where *proportions* are concerned the standard error of the difference= $\sqrt{\frac{p_1q_1}{n_1}+\frac{p_2q_2}{n_2}}$.

An alternative approach is to use p and q, representing the combined proportions in the two samples, when the standard error= $\sqrt{pq\left(\frac{1}{n_1}+\frac{1}{n_2}\right)}$.

### 9.5.1 WORKED EXAMPLE

*Question*

The average weight of a product has been established as 6·205 lb. with a standard deviation of 0·084 lb. Current production shows an average of 6·210 lb. from a sample of 1,600.

*a.* Has the average weight changed significantly?

*b.* How large a sample would be needed to establish the significance of a 0·2% change in the average weight?

*Answer*

*a.* The standard error= $\frac{0{\cdot}084}{\sqrt{1600}}$=0·0021 lb.

Weights may be above or below the established average so we are interested in the two tails of the normal distribution. At the 5% level of significance we should expect a sampling variation of 2×s.e.= 2×0·0021=0·0042 lb.

The actual variation of the sample mean from the assumed population mean is 6·210−6·205=0·005 lb.

Since this is in excess of 0·0042 it must be regarded as significant at the 5% level.

At the 1% level the expected sampling variation would be 2·58×s.e.= 2·58×0·0021=0·00542. The actual variation of 0·005 is less than this so it is not significant at this level.

*b.* A 0·2% change in the average weight=6·205×$\frac{0{\cdot}2}{100}$=0·01241. For this to be significant at the 5% level 2×s.e. must be equal to this.

If there are n items then 2×$\frac{0{\cdot}084}{\sqrt{n}}$=0·01241.

Solving for n gives a value of about 166 items needed if the required degree of change is to be established.

### 9.5.2 WORKED EXAMPLE

*Question*

Two samples of 1,000 items each are taken yielding the following means and standard deviations: $\bar{x}_1=1{\cdot}54$, $\bar{x}_2=1{\cdot}61$, $s_1=1{\cdot}01$, $s_2=0{\cdot}95$. [These are the values obtained in worked example 6·11, relating to the chain store. Here it is assumed that a sample of given size has been taken.]

*a.* Is the difference between the means significant?

*b.* What sample size would just make this difference significant?

*c.* Is it likely that the first sample could have come from a population having a mean of 1·5?

*d.* What sample size would make this sample value of 1·54 significantly different from a population mean of 1·5?

*Answer*

*a.* The standard error of the difference between means

$$=\sqrt{\frac{s^2_1}{n_1}+\frac{s^2_2}{n_2}}=\sqrt{\frac{1{\cdot}01^2}{1000}+\frac{0{\cdot}95^2}{1000}}=0{\cdot}044$$

The difference between the means=1·61−1·54=0·07, which is $\frac{0{\cdot}07}{0{\cdot}044}=1{\cdot}6$ times the s.e. This is not significant at the 5% level, since it is less than 2.

*b.* For the difference in the means to be significant at the 5% level $\frac{\text{difference}}{\text{s.e.}}=2$. If the sample size in each case is assumed to be n then the s.e.= $\sqrt{\frac{1{\cdot}01^2}{n}+\frac{0{\cdot}95^2}{n}}$.

The difference in the means is 0·07. Substitution yields n=1,600.

*c.* For the first sample the s.e.$=\frac{1{\cdot}01}{\sqrt{1000}}=0{\cdot}032$. Twice this gives 0·064. The difference of the sample from the assumed mean=1·54−1·5=0·04. Since this is less than twice the s.e. the sample may well have come from the population.

*d.* For the difference to be significant 0·04=2×s.e. Thus s.e.$=0{\cdot}02=\frac{1{\cdot}01}{\sqrt{n}}$, from which n=2,550.

## 9.6 PRACTICE QUESTIONS

1. Discuss the validity of opinion polls carried out by magazines which invite their readers to send them a postcard or to complete a questionnaire.
2. Give a critical account of the relevance of sampling in your own particular field of study.
3. Explain the following terms: *a.* quota sampling, *b.* activity sampling, *c.* multi-stage sampling.
4. A sample of 100 adult males yields an average height of 68·7 inches. Is there a significant difference from those of qu. 1, section 8.6?
5. In an election there are only two candidates. A survey of electors' voting intentions shows that, of those who intend to vote (numbering 900 in the sample), 410 support one candidate and 490 the other. Can the second be confident of victory?
6. Two factories are comparing their methods for producing the same items. Each takes a sample of 400 items. The times for a particular

stage of production are found to average 2·2 and 2·4 minutes, with standard deviations of 0·5 and 0·7 minutes. Do the times differ significantly?

7. In qu. 5 how many voting intentions would be needed to establish confidence that a 2% difference would result in a victory for a particular candidate?

8. In qu. 6 what size of sample would be needed to establish the significance of a difference between the means of 0·05 minutes?

# 10

# Statistical Quality Control (S.Q.C.)

## 10.1 INTRODUCTION

Statistical Quality Control refers to techniques for assessing and controlling the quality of output and purchases, generally on the basis of samples taken from the current production or from deliveries.

Inspection has always been concerned with separating the bad from the good output. S.Q.C. goes further than this in that it attempts rapidly to assess any significant changes that may be taking place in the production process with a view to remedying these so that the process may be kept in control.

The four most important techniques are:

1. *Measurement Control* using average and range charts.
2. *Cusum* (Cumulative sum) techniques, where the cumulative effects of departures from some standard are measured.
3. *Acceptance Sampling* where sampling is used to check the proportion of defective items in batches produced or delivered.
4. *Percentage Defective Control*, using a chart.

## 10.2 MEASUREMENT CONTROL

This is probably the best established of the S.Q.C. systems. If successive items produced by an automatic machine are measured it is found that no two are alike. The variations are due to:

a. consistent forces which affect the average dimension of a large number of units and which are due to the setting of the machine.

b. random forces which result in the measurement of actual items being dispersed about this average in a way that cannot be predicted for any particular item. These forces depend upon the precision of the machine governed by factors such as the state of its bearings.

This is precisely the situation that arises in the normal curve of error. The production average is the mean of a large number of items and the variability can be measured by the standard deviation. Two charts are used to keep track of the evolving situation.

a. An average chart on which the mean is indicated and which also shows lines drawn at $\pm\frac{3\sigma}{\sqrt{n}}$ and $\pm\frac{2\sigma}{\sqrt{n}}$ limits. As samples are taken their means are plotted on this chart and any falling outside the outer limits indicate the probability that the population is no longer the same or that the mean has altered. Points falling between the inner and the outer limits indicate that the mean may have changed and are used as a warning.

b. Below this chart can be plotted at the same time the range of each sample, and limits can be established for this, so that we keep a check on whether the variability of the machine has altered significantly. Diagram 10.1 shows such a chart based on the data of table 10.1.

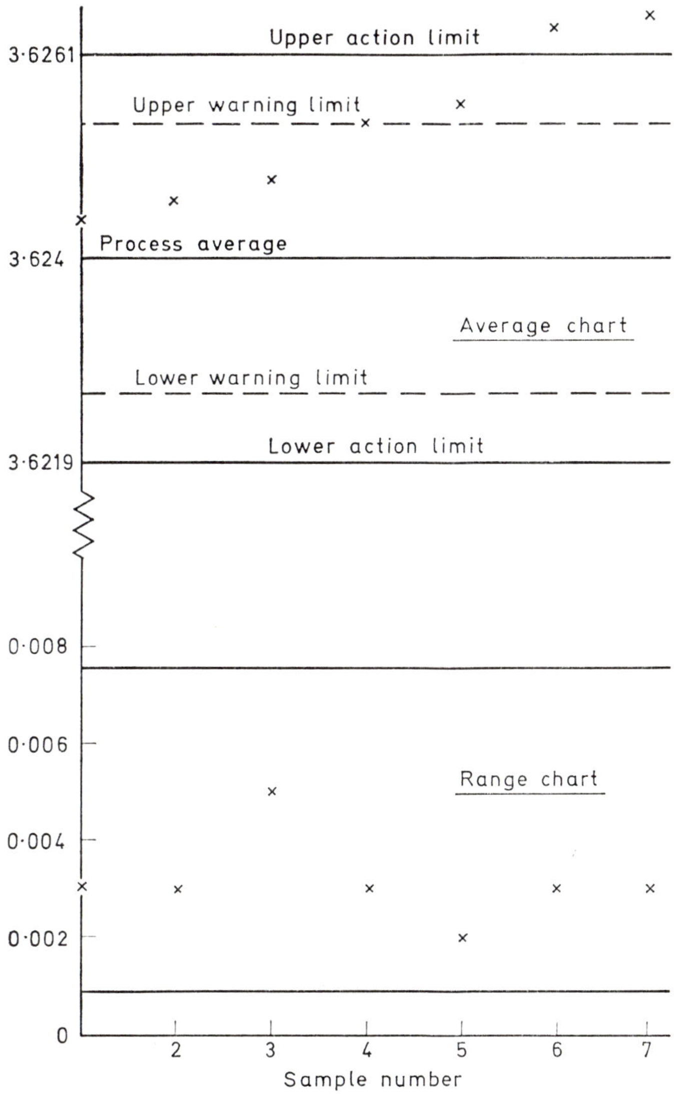


Diagram 10.1. *Average and range charts for measurement control*

The first requirement is to establish the overall average, which is done by taking, say, a hundred items and measuring them. The control limits would be cumbersome to fix using the ordinary techniques used for calculating the standard deviation and we rely on the fact that the standard deviation and the mean range (often shown as $\overline{w}$) of a number of samples are connected. Tables give multiplying factors to enable the control lines on both charts to be fixed.

In the example the first 25 items (25 is really too small a number but is used for purposes of illustration) have been divided into samples of 5 each. The ranges are as shown and the mean range equals $\frac{0{\cdot}018}{5}$ =0·0036. For samples of 5 items tables give the average chart multiplier as 0·59 and the range chart multipliers as 2·09 and 0·25. The outer limits of the average chart are thus at 0·59×0·0036=0·002124 on each side of the average. This average is the total of these 25 items divided by 25, giving 3·62412.

Table 10.1. *Dimensions of 12 samples of 5 items*

| Sample | 1 | 2 | 3 | 4 | 5 | 6 |
|---|---|---|---|---|---|---|
| | 3·624 | 3·626 | 3·625 | 3·624 | 3·625 | 3·624 |
| | 3·623 | 3·625 | 3·622 | 3·626 | 3·623 | 3·623 |
| | 3·624 | 3·625 | 3·624 | 3·624 | 3·625 | 3·625 |
| | 3·624 | 3·624 | 3·623 | 3·624 | 3·624 | 3·626 |
| | 3·627 | 3·620 | 3·625 | 3·625 | 3·622 | 3·624 |
| Total | 18·122 | 18·120 | 18·119 | 18·123 | 18·119 | 18·122 |
| Mean | 3·6244 | 3·624 | 3·6238 | 3·6246 | 3·6238 | 3·6244 |
| Range | 0·004 | 0·006 | 0·003 | 0·002 | 0·003 | 0·003 |
| | 7 | 8 | 9 | 10 | 11 | 12 |
| | 3·623 | 3·625 | 3·624 | 3·625 | 3·626 | 3·627 |
| | 3·625 | 3·627 | 3·627 | 3·627 | 3·625 | 3·625 |
| | 3·626 | 3·624 | 3·626 | 3·625 | 3·627 | 3·626 |
| | 3·624 | 3·622 | 3·625 | 3·626 | 3·626 | 3·628 |
| | 3·625 | 3·626 | 3·625 | 3·625 | 3·628 | 3·627 |
| Total | 18·123 | 18·124 | 18·127 | 18·128 | 18·132 | 18·133 |
| Mean | 3·6246 | 3·6248 | 3·6254 | 3·6256 | 3·6264 | 3·6266 |
| Range | 0·003 | 0·005 | 0·003 | 0·002 | 0·003 | 0·003 |

The upper limit for the range chart is at 2·09×0·0036=0·007524 and the lower limit is at 0·25×0·0036=0·0009.

It should be emphasised that different multiplying constants apply to different sample sizes.

The plots, starting with the sixth sample, show that the process is tending to get out of control at about the fourth plot and requires adjustment after the sixth plot. All plots on the range chart are within the control lines so the precision of the machine has not altered.

The charts give a picture of what is happening to output and if points appear to be drifting in a particular direction we may forecast well in advance when the process will get out of control and plan accordingly. Tool wear will result in an outside diameter dimension gradually increasing, for example.

This control system is interesting as it relies entirely on the characteristics of the production system, not on the design dimension laid down by the drawing office. If the accuracy required by the design is greater than the machine is capable of producing there will inevitably be a high percentage of scrap, which may be calculated using the normal curve tables. On the other hand, if the machine is very accurate compared with the design tolerances we may more or less leave it alone, doing much-reduced sampling as an occasional check.

## 10.3 CUSUM TECHNIQUES

These are simpler to apply than the system just outlined and are claimed to emphasise very quickly any significant changes in the data being studied. Diagram 10.2 shows the cusum chart for the data of table 10.1. Calculations

Table 10.2. *Cusum calculation for data of table 10.1*

A constant of 3·624 has been deducted from each of the 12 sample means and the results multiplied by 1,000 for ease of handling.

| Sample | 1 | 2 | 3 | 4 | 5 | 6 | 7 | 8 | 9 | 10 | 11 | 12 |
|---|---|---|---|---|---|---|---|---|---|---|---|---|
| Devn | +0·4 | 0·0 | −0·2 | +0·6 | −0·2 | +0·4 | +0·6 | +0·8 | +1·4 | +1·6 | +2·4 | +2·6 |
| Cusum | 0·4 | 0·4 | 0·2 | 0·8 | 0·6 | 1·0 | 1·6 | 2·4 | 3·8 | 5·4 | 7·8 | 10·4 |

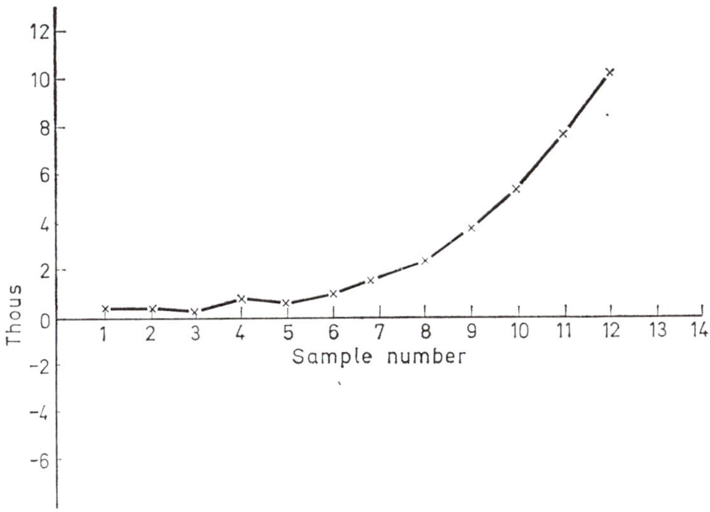


Diagram 10.2. *Cusum chart*

are in table 10.2. The idea is to subtract some constant, approximately equalling the average, from each of the items and then to accumulate the differences. If the data is oscillating about this constant the resulting plus and minus values will tend to cancel out. However, if the data begins to depart from this constant the cumulative effect will be observable in a change in the gradient of the cusum curve, and this is what we look for. It is the slope of the curve not its vertical height that is important.

## 10.4 ACCEPTANCE SAMPLING

The advantage claimed for this technique is that it requires only the ability to differentiate between good and bad items without necessarily making actual measurements. *Go* and *not-go gauges* can be used, probably giving a saving in inspection costs, following from increased speed and less skilled labour.

The aim is to sample lots or batches and to ensure that it is improbable that these batches will contain more than a pre-determined percentage of defectives.

The basis of this technique is the Poisson distribution (see section 8.2). From this can be derived an *operating characteristic*, the general shape of which is shown in diagram 10.3. A curve of this kind can be drawn for different sample sizes (n), and for each number of defectives (c) permitted to appear in the sample without the lot being rejected. Using the figures of section 8·2 the cumulative probability of a sample of 100 items containing no defectives is 0·3679, containing up to one defective is 0·7358, containing up to two defectives is 0·9198. Each of these values would be a point on a separate operating characteristic (for c=0, 1 and 2) located above a value of 1 per cent defective on the horizontal scale. Similar calculations for different percentages in the lot give a series of points that determine the position of the operating characteristics.

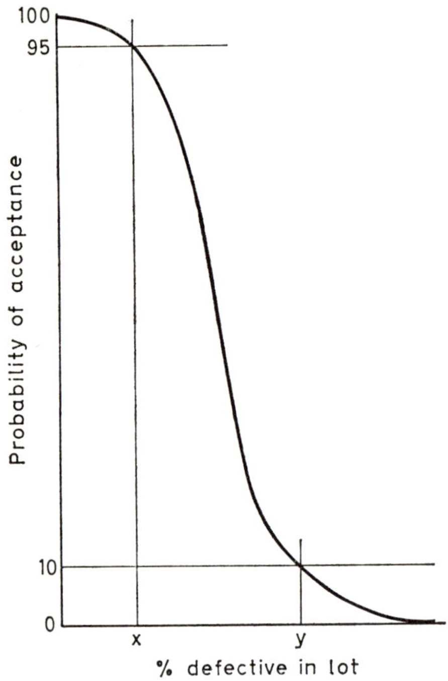


Diagram 10.3.
*Operating characteristics for acceptance sampling*

Tables are available which give the average percentage of defectives being produced by the process (which must be estimated) and which list the lot or batch sizes. Against each batch size is shown the size of sample which should be taken (n) and the maximum permitted number of defects that can occur in this (c) for the batch to be passed as containing not more than the predetermined percentage of defectives. If the number of defectives exceeds c, we must inspect the whole batch or take a larger sample. We can also get an indication of whether the process average has altered (see section 10.5).

More sophisticated techniques work on *double* sampling and *sequential* sampling. The idea is to choose a small sample to start with and hope that this will discriminate between good and bad batches. If it does not we take another sample or several samples, successively, until we can discriminate.

Where the evidence indicates the satisfactory nature of the average quality of the batches going forward for inspection, and where this exceeds the quality level being aimed at, we may use *reduced sampling*.

Diagram 10.3 has lines drawn at the 10% and 95% probability levels. These are often referred to as *consumer's risk* and *producer's risk*, respectively. The former is the risk that the inspection system will miss a batch with a higher than stated percentage of defectives, and obviously the recipient or consumer suffers. The latter is the risk that a satisfactory batch will be rejected and here the producer suffers. Obviously the percentage defective at which these occur (x and y) should be close together, since the acceptance or rejection of values in between them is uncertain. Thus an operating characteristic falling sharply from a high to a low level is to be aimed at. Point x is sometimes called the *acceptable quality level* (A.Q.L.), and point y is termed the *lot tolerance per cent defective* (L.T.P.D.)

The effect of sampling is to remove defective items which are found from the output, and the percentage defectives left after inspection is termed the *average outgoing quality*, (A.O.Q.)

This sampling inspection pays off best when batches are large and the sample required is relatively small. It must be emphasised that it is the absolute sample size that is important, not the percentage this represents of the output.

## 10.5 PERCENTAGE DEFECTIVE CONTROL

This is really similar to measurement control except that, instead of using the average dimension of samples, we work on the percentage defective.

The average level of errors (p%) is first determined by taking a large number (N) of items. The value of the standard deviation is obtained from the formula $\sqrt{\frac{p(100-p)}{n}}$, in which n is to be the sample size. Control limits are chosen, say, at 3 times this value on each side of p and a chart can be immediately set up on which the percentage defectives are plotted. Points outside the control lines indicate the probability that the error conditions have changed.

An alternative is to work on the number of errors, when the standard error will be $\sqrt{np(100-p)}$. This makes for easier plotting, since the percentage does not have to be calculated each time, only the actual number of errors that occur.

This system is often recommended for office work where it can cope with errors such as those arising in typing and calculating. It can also be used in conjunction with acceptance sampling.

## 10.6 WORKED EXAMPLE

*Question*

In a factory a random check of 10% of the output is made and, if no faulty items are found, the particular batch is passed. Comment on this procedure and suggest a better solution.

*Answer*

No sampling method can guarantee to pick out all defectives that might occur. The question does not give details of the firm's attitude towards a permitted level of defectives, if any.

In sampling the percentage of the population that is taken is not the key factor. What matters is whether the actual number of items is sufficient to yield the information we require.

If batches consisted of only 10 items then a 10% sample would select only one item. If the batch were 1,000 items the sample would contain 100.

If the batch has, say, 20% defectives then the chance of one of the defectives turning up in a sample of 1 is 0·2, and the chance of the sample being clear is 0·8. Thus 80% of the time we should miss even this very high level of defectives.

With a sample of 100 the chance of no defectives turning up in any batch is $(0{\cdot}8)^{100}$, which is very small indeed, so that most of the time the batches would be rejected and perhaps tested further.

A satisfactory approach to this sort of problem is afforded by *acceptance sampling* [then give as full an explanation as time permits, based on 10.4].

## 10.7 PRACTICE QUESTIONS

1. 'In the past inspection was concerned simply with separating good output from bad. With quality control the emphasis is placed on monitoring the production process and even forecasting when intervention will be required.' Discuss this statement.
2. Compare and contrast measurement control and acceptance sampling.
3. Discuss the application of statistical quality control to audit procedures in an office.
4. What is the implication of the statement that a process is 'out of control'?
5. Statistical quality control is said to be especially economical when long production runs or large batches are concerned. Explain why this is so.

# 11

# Straight Lines and Regression

## 11.1.1 THE STRAIGHT LINE AND ITS FORMULA

Many practical situations may be simply expressed by a straight line formula of the type

$$y=A+Bx.$$

In this formula A and B are constants for the particular situation being considered, while x and y are variables, that is they can take on many values.

Diagram 11.1 shows two straight line graphs which can be represented by

$$y=7x$$

and $y=10{,}000+5x$.

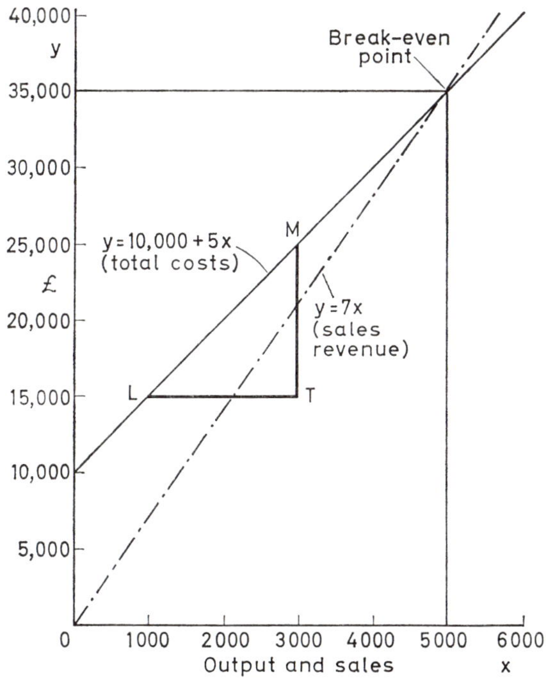


Diagram 11.1. *Straight lines.* Simple break-even chart

In the first case there is no A constant, and B is 7. It will be seen that this represents the gradient of the graph or the amount by which it rises on the y axis for every unit increase on the x axis. This graph passes through zero on both axes (a point usually called the *origin* of the graph) and this will always be the case where there is no fixed amount, such as A, in the equation.

In the second case A has the value of 10,000, so the graph starts at this point on the y axis (called the *intercept* on the y axis) and then rises at the rate of 5 units of y for every unit of x.

The value plotted horizontally is usually the *independent variable*, i.e. the one whose values we have a choice in selecting, while vertically we normally plot the *dependent variable*, i.e. the one obtained from the equation.

The following should also be noted:

a. Where no term involving x occurs (B=0) then the line neither rises nor falls but lies parallel to the x axis and A units from it (e.g. y=A, y=5, y=10 etc.).

b. Where y does not appear, e.g. x=6, the line rises vertically over the value of 6 on the x axis.

c. A negative value for B indicates that the graph is falling instead of rising.

d. Any point on the graph may be referred to by quoting, in brackets, first its x value, then its y value. The point of intersection of the two lines would be (5,000, 35,000).

Many readers may recognise that the graphs shown amount to a simple *break-even chart* as used by accountants. The line through the origin would be the sales revenue from selling x units at £7 each. The other line would represent the costs of producing the volume of output shown by x. The intercept shows the *fixed costs* (overheads) and the gradient gives the *variable costs* (materials, labour etc.). Where the total costs and the sales revenue are equal we have the break-even point, where neither profit nor loss is made and to the right of which the business must try to operate.

If we deduct the cost from the sales revenue we have an estimate of the profit (p) for any output, i.e.

$$p=7x-(10{,}000+5x)=2x-10{,}000.$$

Cost accountants will recognise that 2x is the *contribution* of marginal costing (the excess of the price of each item over its variable cost).

It is important to recognise that a formula has a practical interpretation and is not just an abstraction.

### 11.1.2 FITTING A STRAIGHT LINE THROUGH TWO POINTS

We need only two points to fix completely the position of a straight line. From these point the values of the constants A and B can easily be determined.

Suppose that we do not know the formula of the cost line (diagram 11.1) but have information about two points on it, L (1,000, 15,000) and M (3,000, 25,000).

A simple approach is to draw a graph linking these points. Where this cuts the y axis (N) gives the value of the constant A (10,000). The gradient, B, can be found by taking any two convenient points on the line (L and M themselves will do in the example given) and dividing the length MT (in scale units) by the length LT (again scaled). This gives the gradient as $\frac{10{,}000}{2{,}000}=5$, which is the value of B.

A graph is not needed. The gradient can be first calculated, as above, from the two points provided. Substitution in the formula y=A+Bx gives y=A+5x, and the further use of either of the pairs of values for x and y will yield A. Thus, taking the values for point L, we have, $15{,}000=A+5\times 1{,}000$, from which we find that A must be 10,000.

A simple formula gives the same result. If the known points have co-ordinates $(x_1, y_1)$ and $(x_2, y_2)$, respectively, then the line connecting them can be obtained from

$$\frac{x-x_1}{x_2-x_1}=\frac{y-y_1}{y_2-y_1}.$$

In the example for points L and M we have

$$\frac{x-1{,}000}{3{,}000-1{,}000}=\frac{y-15{,}000}{25{,}000-15{,}000}$$

from which $y=10{,}000+5x$.

## 11.2.1 REGRESSION

This chapter has so far been concerned with the very simple algebraic approach to the straight line. A knowledge of this is required to understand what follows and to cope with many problems in statistics.

The statistical approach to the straight line is concerned with cases where there are more than just the two points needed to fix such a line and the available values do not lie exactly on any line. This often occurs with practical experimental and economic data.

Table 11.1 is a simple illustration. It shows how ten workers, designated by the letters A to J, scored in a selection test (x) and how their work performances were subsequently rated (y). If the test is satisfactory we should expect some relationship to be apparent between the two sets of scores. A straight line might reasonably represent this relationship.

Table 11.1. *Selection test scores*

| Worker | A | B | C | D | E | F | G | H | I | J |
|---|---|---|---|---|---|---|---|---|---|---|
| Test % (x) | 85 | 82 | 76 | 70 | 66 | 60 | 60 | 58 | 56 | 55 |
| Performance % (y) | 77 | 86 | 90 | 75 | 80 | 70 | 74 | 65 | 65 | 65 |

These points (one for each pair of scores) have been shown in diagram 11.2. This is called a *scatter diagram* and will be referred to later in connection with correlation (section 12.2).

The line connecting y and x in cases such as this is called the *line of regression of y on x*. Note that we do not attempt to connect up all the points. The regression line shows the general connection between y and x and none of the actual points need lie on it. Ignore the broken line for the moment.

The simplest approach to finding the equation of the regression line is to estimate its position on the scatter diagram (moving a ruler about until it seems to fit) and then to use the method given above to estimate the constants.

The method of least squares next referred is a much superior technique, yielding the same results no matter who applies it, and eliminating the subjective factor involved in estimation. Other techniques especially suited to time series, where the horizontal axis usually refers to equal time intervals, are dealt with at 14.2.

## 11.2.2 THE METHOD OF LEAST SQUARES

Before we can fit a line to the data of the type shown in the scatter diagram we have to decide on a criterion of best fit. Obviously the line must be as close to the points as is possible, but what does this mean? If we interpret it to imply that the average of the distances of the points from the line should be as small as possible the same problem is encountered as in the case of the mean deviation: positive and negative variations tend to cancel each other, no matter how large they are.

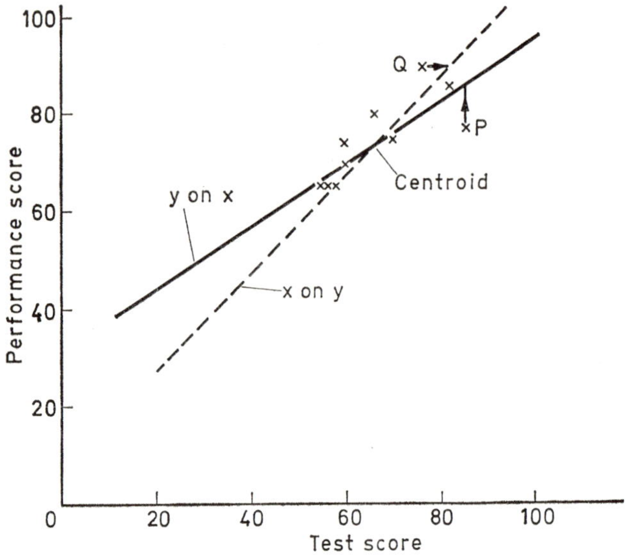


Diagram 11.2. *Scatter diagram with regression lines*

The most usual solution is to select a line such that the sum of the squares of the distances of the points from the corresponding points on the line is a minimum. This implies that, for the line of regression of y on x, and for any given value of x, we take the difference between the ordinate (y value) of the actual point and the ordinate corresponding to this on the line.

One such deviation, for point P, is shown by the arrowed line in the scatter diagram. Notice that this is not the normal way of interpreting the distance from a point to a line, which is usually taken to be the shortest distance, or perpendicular, and would be the length of a line from P cutting the regression line at right angles.

It will be seen that, if the scatter diagram is viewed from the y axis as base, the deviations from the line will be parallel to the x axis and will differ in amount from those just considered. It is because of this that there are two possible regression lines for the same data. The broken line shows the line of regression of x on y. The deviation of a point Q from it is marked.

The standard approach to the problem of fitting such a line which will be of the form

$$y=A+Bx$$

is to work from what are called the normal equations:

$$\Sigma y=An+B\Sigma x \quad \ldots\ldots\ldots\ldots\ldots\ldots \quad (1)$$

$$\text{and } \Sigma xy=A\Sigma x+B\Sigma x^2 \quad \ldots\ldots\ldots\ldots\ldots\ldots \quad (2).$$

We calculate the values of $\Sigma x$, $\Sigma y$, $\Sigma x^2$ and $\Sigma xy$ from the tabulated data, substitute in these two equations, and then solve them simultaneously to give the required values of A and B. It can be shown that these values will give a line such that the sum of the squares of deviations of the points from it will be minimum.

Such a calculation is outlined in table 11.2.

Table 11.2. *Using the method of least squares to fit a regression line to the test data*

| Worker | x | y | $x^2$ | xy | $y^2$ |
|---|---|---|---|---|---|
| A | 85 | 77 | 7225 | 6545 | 5929 |
| B | 82 | 86 | 6724 | 7052 | 7396 |
| C | 76 | 90 | 5776 | 6840 | 8100 |
| D | 70 | 75 | 4900 | 5250 | 5625 |
| E | 66 | 80 | 4356 | 5280 | 6400 |
| F | 60 | 70 | 3600 | 4200 | 4900 |
| G | 60 | 74 | 3600 | 4440 | 5476 |
| H | 58 | 65 | 3364 | 3770 | 4225 |
| I | 56 | 65 | 3136 | 3640 | 4225 |
| J | 55 | 65 | 3025 | 3575 | 4225 |
| | 668 | 747 | 45706 | 50592 | 56501 |

The $y^2$ column is not needed for the line of regression of y on x, but is included to enable the line of regression of x on y to be calculated as well.

The normal equations become

$$747 = 10A + 668B \quad \ldots\ldots\ldots \quad (1)$$
$$50{,}592 = 668A + 45{,}706B \quad \ldots\ldots\ldots \quad (2)$$

Divide (1) by 10 and (2) by 668 to get

$$74{\cdot}7 = A + 66{\cdot}8B \quad \ldots\ldots\ldots \quad (3)$$
$$75{\cdot}74 = A + 68{\cdot}42B \quad \ldots\ldots\ldots \quad (4)$$

By subtraction

$$1{\cdot}04 = 1{\cdot}62B$$

Thus $B = 0{\cdot}64$

and, by substitution in (3)

$$A = 32.$$

The regression equation of y on x is thus

$$y = 32 + 0{\cdot}64x.$$

To find the line of regression of x on y we should use the normal equations

$$\Sigma x = An + B\Sigma y$$
and $\Sigma yx = A\Sigma y + B\Sigma y^2$.

Substitution of the values from the table and solution of the simultaneous equations yields

$$x = 0{\cdot}99y - 7{\cdot}1.$$

It must be emphasised that the two regression equations are not the same, and one cannot be obtained directly from the other.

The calculations for the regression equations are, perhaps, simplified by working from an assumed origin and then adjusting the equations. If we take an origin at (67, 75) and call the adjusted data X and Y, where $X = x - 67$ and $Y = y - 75$, then the calculations are as in table 11.3 from which $\Sigma X = -2$, $\Sigma Y = -3$, $\Sigma X^2 = 1084$ and $\Sigma XY = 693$.

The normal equations become:

$$-3=10A-2B$$
$$693=-2A+1084B$$

whence $B=0{\cdot}64$

$$A=-0{\cdot}1$$

giving the regression equation as

$$Y=-0{\cdot}1+0{\cdot}64X.$$

This is based on an origin located at (67, 75), as far as the original values are concerned. Thus $X=x-67$ and $Y=y-75$,

giving $(y-75)=-0{\cdot}1+0{\cdot}64\,(x-67)$, simplifying to

$y=32+0{\cdot}64x$, in terms of the original values.

This process is often called *changing the origin* of an equation.

Table 11.3

| | X | Y | $X^2$ | XY | $Y^2$ |
|---|---|---|---|---|---|
| A | 18 | 2 | 324 | 36 | 4 |
| B | 15 | 11 | 225 | 165 | 121 |
| C | 9 | 15 | 81 | 135 | 225 |
| D | 3 | 0 | 9 | 0 | 0 |
| E | −1 | 5 | 1 | −5 | 25 |
| F | −7 | −5 | 49 | 35 | 25 |
| G | −7 | −1 | 49 | 7 | 1 |
| H | −9 | −10 | 81 | 90 | 100 |
| I | −11 | −10 | 121 | 110 | 100 |
| J | −12 | −10 | 144 | 120 | 100 |
| | 45 | 33 | 1084 | 698 | 701 |
| | −47 | −36 | — | −5 | — |
| | −2 | −3 | | 693 | |

Both regression lines pass through the *averages of the x and y values* and this point is sometimes called the centre of gravity, or centroid, of the data.

This fact can afford an alternative to the method of solving the normal equations. It provides one point on the line. The gradient can be found from the general solution of the normal equations for B, which gives

$$B=\frac{\Sigma y\Sigma x-n\Sigma xy}{(\Sigma x)^2-n\Sigma x^2}.$$

Substitution, as in 11.1.2 then yields the value of A. An example is given at the end of this chapter.

An examination of the formula for B, just quoted, shows that, if $\Sigma x=0$, this simplifies to $\frac{\Sigma xy}{\Sigma x^2}$. This can be used *where the x values increase by equal steps* so that they can be converted into + and − values about a central point. The average of these values will be zero. This is especially important in the case of time series, where x represents dates, increasing, perhaps, by yearly or monthly intervals. Examples are given in chapter 14, but it should not be forgotten that this technique is not restricted to time series.

## 11.3 WORKED EXAMPLE

*Question*

Fit a straight line by the method of least squares to the following data

| | | | | | |
|---|---|---|---|---|---|
| x | 0 | 2 | 3 | 5 | 8 |
| y | 11 | 33 | 40 | 60 | 91 |

*Answer*

[The method followed is to find the gradient, and then to use the fact that the line will pass through the means of the x and y values. Other methods give the same result.]

Table 11.4

| x | y | xy | $x^2$ |
|---|---|---|---|
| 0 | 11 | 0 | 0 |
| 2 | 33 | 66 | 4 |
| 3 | 40 | 120 | 9 |
| 5 | 60 | 300 | 25 |
| 8 | 91 | 728 | 64 |
| 18 | 235 | 1214 | 102 |

From table 11.4 we have:

$$\bar{x}=\frac{18}{5}=3{\cdot}6 \text{ and } \bar{y}=\frac{235}{5}=47,$$

$$\Sigma x=18,\ \Sigma y=235,\ \Sigma xy=1214,\ \Sigma x^2=102$$

Substitution in the formula for the gradient

$$B=\frac{\Sigma y\Sigma x-n\Sigma xy}{(\Sigma x)^2-n\Sigma x^2}$$

gives $B=\dfrac{235\times 18-5\times 1214}{18^2-5\times 102}=9{\cdot}9.$

We know that this passes through the centroid which is the point (3.6, 47) on the curve $y=A+Bx$, so that

$$47=A+9{\cdot}9\times 3{\cdot}6, \text{ whence } A=11{\cdot}4.$$

The required line is thus

$$y=11{\cdot}4+9{\cdot}9x.$$

## 11.4 PRACTICE QUESTIONS

1. Given the following values of x and y:

| | | | | | | | | | |
|---|---|---|---|---|---|---|---|---|---|
| x | 1 | 3 | 4 | 7 | 10 | 11 | 12 | 15 | 20 |
| y | 30 | 39 | 40 | 51 | 62 | 63 | 71 | 82 | 94 |

   *a.* Draw a scatter diagram and show a likely regression line.

   *b.* Find the equation of the line you have drawn.

   *c.* Find the equation of the least squares regression line of y on x.

   *d.* Use the equation to calculate the value of y when x is 18.

2. From the data of qu. 1 find the equation of the least squares regression line of x on y.

3. Two factories produce identical products. Over similar periods of time the overheads of factory A are £5,000 and its direct costs for each unit are £25, while for factory B the overheads are double this but the direct costs are only half.

   *a.* Show the total cost equations of the two factories.

   *b.* At what level of production will factory B show a cost advantage over factory A?

4. *a.* Explain why there are two regression lines in the case of paired values of items.

   *b.* At what point do these two regression lines cross?

   *c.* If the regression lines coincide what does this imply?

# 12

# Correlation

## 12.1 SIMPLE CORRELATION AND STATISTICAL RELATIONSHIPS

*Simple correlation* is concerned with the situation when two sets of figures appear to move in conjunction.

With a small number of items inspection of the data may indicate whether there seems to be any relationship between the sets of data, and the *scatter diagram*, described below, affords a method of bringing out such a connection. Often we require to go beyond this and to indicate how strong the relationship appears to be. For this purpose some form of *coefficient of correlation* is needed.

Before we deal with simple correlation it must be observed that it takes no account of other factors which may be relevant to the situation. If we wish to take these factors into account then *multiple correlation* may be the technique to use, where the significance of the part played by each in the total result is indicated, or *partial correlation*, where it is desired to show the relationship between the two factors 'other things being equal'.

It should be quite clear that the existence of correlation between two sets of data does not necessarily prove that one is the cause of the other. Correlation indicates a *statistical relationship*. This was emphasised in the early results showing the correlation between heavy smoking of cigarettes and lung cancer. It was necessary to carry out medical research to see if there was any causal connection.

## 12.2 SCATTER DIAGRAMS

A scatter diagram was used to illustrate the selection test data of the previous chapter (diagram 11.2). There is a point for every pair of values and the pattern is examined for any tendency to settle about a straight line.

In the case illustrated high values of y tend to be connected with high values of x, giving a line rising from left to right across the graph. Such correlation is called *positive* or *direct*.

If low values of y are connected with high values of x the pattern of points will tend to cut across the intersection of the axes, falling from left to right. This is termed *negative* or *inverse* correlation.

Often whether the correlation appears to be positive or negative is a matter of how the data are expressed. If it is a matter of accident proneness testing the score may be based on the number of mistakes made in the test, which should tend to correlate positively with the number of accidents subsequently made at work. Alternatively, the number of errors could be subtracted from, say, 100, so that a high score indicated the person's resistance to accident, in which case the later correlation would be negative.

The same considerations would apply to economic data correlated with the percentages of persons unemployed, in the one case, or employed, in the other. Unemployment or employment percentages are simply alternative ways of stating the same facts.

## 12.3 TIME SERIES ASSOCIATION AND LAG

An important consideration in business and the social sciences is whether one series, perhaps occurring over a number of years or months, has movements associated with those of some other series.

We might wish to investigate how bank loans or exports correlate, if at all, with changes in the bank rate, or how the introduction of new management techniques correlate with productivity.

If a time lag is to be expected before one set of changes can be expected to be reflected in the others we must correlate, not the two series as they arise, but one series with the data for the other, separated by the appropriate period of *lag*.

Graphing both series on the same base enables the probable lag to be estimated. It is also a help to plot one series on transparent paper which can be moved over a graph of the other series so that the effect of different estimates of the lag can be appreciated.

## 12.4.1 PEARSON'S COEFFICIENT OF CORRELATION

It will be seen that the calculations for fitting a line by the method of least squares involve the multiplication of x and y, or of the deviations of x and y from their respective means.

By analogy with the calculation of variance we may say that the *covariance* is the result of multiplying together the deviation from their means of each pair of items, adding the results, and dividing by the number of pairs of items involved, i.e.

$$\text{covariance} = \frac{\Sigma(x-\bar{x})\,(y-\bar{y})}{n} = \frac{\Sigma xy}{n} - \bar{x}\bar{y}$$

Obviously, if the two series move together, so that large items of x are associated with large items of y, the covariance will have a positive value.

If, on the other hand, there is inverse correlation, plus and minus deviations will tend to be multiplied, giving a negative value.

Where there is no connection between the two series plus and minus will tend to cancel out with a large number of items.

This property can be used to give a measure of correlation and is the basis of the best known coefficient of correlation, the *product moment coefficient* of correlation (Pearson's). This is denoted by r, defined as

$$r = \frac{\text{covariance}_{xy}}{\sqrt{\text{variance}_x \cdot \text{variance}_y}} = \frac{\text{convariance}_{xy}}{\sigma_x \sigma_y}$$

It can be shown that this measure varies from −1 to +1, the former indicating complete inverse correlation and the latter perfect direct correlation.

Other versions of the formula for r, which may be more convenient for calculation, depending on how the figures are presented, are

$$r=\frac{\Sigma xy-n\bar{x}\bar{y}}{n\sigma_x\sigma_y}$$

$$\text{and } r=\frac{\Sigma(x-\bar{x})\,(y-\bar{y})}{n\sigma_x\sigma_y}$$

If the coefficient for the selection test data is required we can use the totals already calculated for the regression equations (table 11.2).

$$\Sigma xy=50592$$
$$\bar{x}=66{\cdot}8$$
$$\bar{y}=74{\cdot}7$$

$$\sigma_x=\sqrt{\frac{45706}{10}-\left(\frac{668}{10}\right)^2}=10{\cdot}4$$

$$\sigma_y=\sqrt{\frac{56501}{10}-\left(\frac{747}{10}\right)^2}=8{\cdot}37$$

$$r=\frac{50592-10\times 66{\cdot}8\times 74{\cdot}7}{10\times 10{\cdot}4\times 8{\cdot}37}$$
$$=0{\cdot}8$$

Actually, if the regression lines are available (which is unlikely) we may use the fact that the coefficient of correlation is equal to the square root of the product of the gradients of the two regression equations.

In this case

$$r=\sqrt{0{\cdot}64\times 1}$$
$$=\ 0{\cdot}8$$

The significance of this coefficient is discussed at 12.7.

## 12.4.2 CORRELATION TABLES

The example used to illustrate correlation up to now has been ofı two series paired together, so that we have n pairs of values, each of which can be shown on a scatter diagram. Often the practical situation is more like the one shown in the correlation table (table 12.1) where it is possible for each pair of items, indicated by the intersections of rows and columns, to occur more than once, the frequency of occurrence being indicated .by the number in the cell.

Table 12.1. *Correlation table*
*Showing how values of y are associated with those of x*

| y | x— 10 | 11 | 12 | 13 | 14 | 15 | Total |
|---|---|---|---|---|---|---|---|
| 25 | 1 | 1 | | | | | 2 |
| 35 | 2 | 4 | 3 | | | | 9 |
| 45 | | 1 | 4 | 3 | 1 | | 9 |
| 55 | 1 | | 3 | 1 | 2 | | 7 |
| 65 | | | 1 | 1 | | 1 | 3 |
| Total | 4 | 6 | 11 | 5 | 3 | 1 | 30 |

The x and y values can be discrete or grouped, e.g. the y values in the above table might be the centres of groups 20 to 30, 30 to 40 etc.

A simple way of dealing with data of this kind is to calculate the average values of one variable associated with each value of the other, and to plot these two series on a scatter diagram. In other words we can convert the table into a list of paired items by assuming that each column (or each row) can be expressed by its average.

Thus, looking at table 12.1, we see that when $x=10$ there are 4 items in this column. Their average y value is $(25\times1+35\times2+55\times1)\div4=37{\cdot}5$. This gives one point (10, 37·5) on such a scatter diagram. The other 5 points in this case would be obtained from similar treatment of the remaining 5 columns.

Calculation for the full data is shown below. In this case assumed means of 12 and 45 have been taken, which turn out also to be the actual means of the x and y values.

The value of the covariance may be calculated as in table 12.2. In each cell of the table is first placed the product of the deviations for the row and column

Table 12.2. *Calculation of the covariance*

| $d_y$ | $d_x$ – | −2 | −1 | 0 | 1 | 2 | 3 |
|---|---|---|---|---|---|---|---|
| −2 | | 4 (4) | 2 (2) | 0 | | | |
| −1 | | 2 (4) | 1 (4) | | | | |
| 0 | | 0 | 0 | 0 | 0 | 0 | 0 |
| 1 | | −2 (−2) | | | 1 (1) | 2 (4) | |
| 2 | | | | | 2 (2) | | 6 (6) |

$\Sigma f(x-\bar{x})(y-\bar{y})$ = Sum of brackets × working units $(27-2)\times(1\times10)=250$

concerned. These are then multiplied by the frequency of the relevant cells in the original table. These products are placed in brackets in the table shown. All these products are added, taking account of signs, for the whole of the table, to give the value of $\Sigma f(x-\bar{x})(y-\bar{y})$. Division by ≏ gives the covariance.

If we take the cell where $x=10$ and $y=25$ then the deviations from 12 and 45 (in 10s) respectively are −2 and −2. The product of these is +4 which is multiplied by the frequency, 1, to give the 4 in the brackets. For the cell (14, 55) the product of the deviations is $2\times1=2$ which is multiplied by the frequency of 2 to give the 4 in brackets.

Table 12.3. *Calculation of the standard deviations*

| Dev. from 12 ($d_x$) | f | $fd_x$ | $fd_x^2$ | Dev. from 45 in 10s ($d_y$) | f | $fd_y$ | $fd_y^2$ |
|---|---|---|---|---|---|---|---|
| −2 | 4 | −8 | 16 | −2 | 2 | −4 | 8 |
| −1 | 6 | −6 | 6 | −1 | 9 | −9 | 9 |
| 0 | 11 | 0 | 0 | 0 | 9 | 0 | 0 |
| 1 | 5 | 5 | 5 | 1 | 7 | 7 | 7 |
| 2 | 3 | 6 | 12 | 2 | 3 | 6 | 12 |
| 3 | 1 | 3 | 9 | | | | |
| | 30 | −14 | 48 | | 30 | −13 | 36 |
| | — | 14 | — | | | 13 | |
| | | 0 | | | | 0 | |

$\bar{x}=12+0=12$

$\bar{y}=45+0=45$

$$\sigma_x=\sqrt{\frac{48}{30}-(0)^2}=1{\cdot}27$$

$$\sigma_y=10\times\sqrt{\frac{36}{30}-(0)^2}=11$$

The values of the standard deviations of the two sets of values can be found in the normal way, as in table 12.3.

The coefficient of correlation then follows as

$$r=\frac{250}{30\times 1{\cdot}27\times 11}=0{\cdot}6.$$

## 12.5 RANK CORRELATION

The product moment coefficient of correlation explained above often has the disadvantage of involving cumbersome calculation. With paired data an alternative approach is to consider the comparative order or rank of corresponding items in the two series.

The calculation for the selection test data is shown in table 12.4. Whether we call the largest or the smallest the first item does not matter as long as we treat both columns the same way. Where two or more items in a column are the same they are each given the average of the ranks they would occupy.

Table 12.4

| Worker | Rank | | Difference in | $D^2$ |
|---|---|---|---|---|
| | Test | Performance | Rank (D) | |
| A | 1 | 4 | −3 | 9 |
| B | 2 | 2 | 0 | 0 |
| C | 3 | 1 | 2 | 4 |
| D | 4 | 5 | −1 | 1 |
| E | 5 | 3 | 2 | 4 |
| F | 6½ | 7 | −½ | ¼ |
| G | 6½ | 6 | ½ | ¼ |
| H | 8 | 9 | −1 | 1 |
| I | 9 | 9 | 0 | 0 |
| J | 10 | 9 | 1 | 1 |
| | | | 5½ | 20½ |
| | | | −5½ | |

Thus, in the test data F and G are equal, they share 6th and 7th places, so are each ranked 6½. The next item is then ranked 8th, so we finish with the last item, ranked 10th.

The values of $\Sigma D^2$ is then substituted in the formula given below, in which 6 is a constant and n is the number of pairs of items.

Coefficient of rank correlation

$$=1-\frac{6\Sigma D^2}{n\,(n^2-1)}$$

$$=1-\frac{6\times 20\frac{1}{2}}{10\,(100-1)}$$

$$=0{\cdot}88$$

The formula for rank correlation can be derived from that for product moment correlation. It may help in a simple understanding of it to note that, when the two series have exactly the same rankings, $\Sigma D^2=0$ and the coefficient is 1. When the rankings are the reverse in one series to the other $\Sigma D^2=2$ and the coefficient is $-1$. You may care to check this latter point by writing down the numbers from 1 to any chosen value, then pairing them in reverse.

## 12.6 AUTOCORRELATION AND SERIAL CORRELATION

If we have a series of values occurring at fixed time intervals then autocorrelation measures the correlation between pairs of values in the same series that are a fixed number of intervals apart. The time 'distance' is called the *lag* and is often indicated by $k$ (which is a positive integer).

When a time series has a fairly regular up and down movement the coefficient of autocorrelation will be greatest when k represents the wavelength or cycle time for the data. In this case large values will be associated with large, and small with small.

Calculation of such coefficients using different values of k enables the most appropriate lag to be found. A graph of r plotted against k is called a *correlogram.*

The term *serial correlation* is synonymous with autocorrelation, though it is sometimes restricted to sampling estimates of the latter.

The reader may like to use the data of yearly sales totals in table 14.1 to calculate correlation coefficients with lags of 1, 2, 3 etc. years. For $k=5$, for example, the values to be correlated will be the original series side by side with the same series starting 5 items on; thus we have 55 and 69, 51 and 64, 44 and 52 etc.

## 12.7 SIGNIFICANCE OF CORRELATION

We know that the coefficient of correlation can take on values from $-1$ to $+1$. Obviously, the closer to unity, the higher is the degree of correlation. How low can this coefficient be and still indicate a significant degree of correlation?

The significance depends on the number of pairs of items in the sample. If, for example, we have a sample of one pair, such as might be obtained by tossing two coins, we can only have the results HH, HT, TH, TT. Of these, HH and TT imply perfect direct correlation and HT and TH imply perfect inverse correlation. Clearly which we obtain depends entirely on chance and the particular result has no significance. The same sort of argument applies to samples of the more usual kind. With a very small number some degree of correlation is likely to be present, purely by chance.

For samples of 10 or more items a simple test is to calculate the standard error of distribution of $r$, which is $\frac{1}{\sqrt{n-1}}$, or, approximately, $\frac{1}{\sqrt{n}}$ if n is large. If r is more than twice this it may be regarded as significant at the 5% level of confidence.

With the selection test data already used n is 10, so the standard error is $\frac{1}{\sqrt{9}}=0{\cdot}33$. Twice this is 0·67. Since the calculated value of r (0·8) exceeds this, we may regard it as showing a significant correlation between the two sets of data.

It should be noted that, if we have a very large sample, the standard error will be very small, so that a low value of r may be significant. We must distinguish between a value of r that is not significant, and a low value that is significant but indicates only a weak degree of association.

When a large sample shows a low coefficient of correlation the standard error is generally taken to be $\frac{1-r^2}{\sqrt{n}}$.

The best test of whether r differs significantly from zero for small samples is the *t-test*, and that for checking if two values of r from different samples differ significantly is the *z-transformation*. These are dealt with in chapter 20.

## 12.8 WORKED EXAMPLE

*Question*

Table 12.5 (a) summarises an attempt to find if there is any statistical connection between the average money values of the monthly orders of 30 customers for two main products, x and y, of a manufacturer. The row and column headings represent these values, in £000's, grouped for convenience. The figures in the table are the frequencies of occurrence of combinations of values of x and y.

Calculate the coefficient of correlation and comment on its significance.

Table 12.5

(a)

| x | y—0 — 10 | — 20 | — 30 | — 40 | — 50 |
|---|---|---|---|---|---|
| 0—10 | 1 | 1 | — | — | — |
| —20 | — | 2 | 3 | — | 2 |
| —30 | — | — | 4 | 3 | 4 |
| —40 | — | 1 | — | 5 | 3 |
| —50 | — | — | — | 1 | — |

(b)

| $d_x$ | $d_y$ | f | $fd_x$ | $fd_y$ | $fd^2_x$ | $fd^2_y$ | $fd_xd_y$ |
|---|---|---|---|---|---|---|---|
| −2 | −2 | 1 | −2 | −2 | 4 | 4 | 4 |
| −2 | −1 | 1 | −2 | −1 | 4 | 1 | 2 |
| −1 | −1 | 2 | −2 | −2 | 2 | 2 | 2 |
| −1 | 0 | 3 | −3 | 0 | 3 | 0 | 0 |
| −1 | 2 | 2 | −2 | 4 | 2 | 8 | −4 |
| 0 | 0 | 4 | 0 | 0 | 0 | 0 | 0 |
| 0 | 1 | 3 | 0 | 3 | 0 | 3 | 0 |
| 0 | 2 | 4 | 0 | 8 | 0 | 16 | 0 |
| 1 | −1 | 1 | 1 | −1 | 1 | 1 | −1 |
| 1 | 1 | 5 | 5 | 5 | 5 | 5 | 5 |
| 1 | 2 | 3 | 3 | 6 | 3 | 12 | 6 |
| 2 | 1 | 1 | 2 | 1 | 4 | 1 | 1 |
| | | 30 | 0 | +21 | 28 | 53 | +15 |

*Answer*

[This is a correlation table and may be dealt with as shown at 12.4.2. With a simple table of this kind it is often easier to convert the data into the form of section 12.4.1, which has been done here using one row for each occupied cell. The centres of the groups have been taken and an assumed mean of 25 used for x and y. In table 12.5 (b) $d_x$ and $d_y$ are deviations from 25 in units of 10. Since we are concerned only with correlation we can work straight through in these units as relative sizes of x and y will not affect the calculation.]

Working in units of 10 from assumed means of 25 we have:

$$s_{d_x} = \sqrt{\frac{28}{30}} = 0{\cdot}966 \text{ (the adjustment is zero)}$$

$$s_{d_y} = \sqrt{\frac{53}{30} - \left(\frac{21}{30}\right)^2} = 1{\cdot}17$$

$$\text{cov}_{d_x d_y} = \frac{15}{30} - \frac{0}{30} \times \frac{21}{30} = 0{\cdot}5$$

$$r = \frac{0{\cdot}5}{0{\cdot}966 \times 1{\cdot}17} = 0{\cdot}44.$$

This seems to be a fairly low coefficient of correlation from a sample that is not very large. Application of the $\frac{1-r^2}{\sqrt{n}}$ criterion of the standard error yields $\frac{1-0{\cdot}44^2}{\sqrt{30}} = 0{\cdot}15$. Since the value of r is three times this, almost, we may assume that the correlation is real and not just due to chance.

[Application of Student's t-test, as in example 20.8.2, confirms this.]

## 12.9 PRACTICE QUESTIONS

1. Calculate the coefficient of correlation between the x and y values of qu. 1 of the previous chapter.

2. Calculate the coefficient of rank correlation between the data in questions 1 and 2 of chapter 14 for the years 1957—1967. Comment on your result.

3. Calculate the coefficient of correlation between s and t from the following table:

| s | t | 2 | 4 | 6 | 8 | 10 |
|---|---|---|---|---|---|---|
| 0—10 | | 2 | 1 | | | |
| 10—20 | | 1 | 6 | 5 | | |
| 20—30 | | 1 | 4 | 12 | 8 | 2 |
| 30—40 | | | 2 | 4 | 10 | 12 |

4. Convert the data of qu. 3 into a form in which it can be shown on a scatter diagram, and draw the latter.

5. Explain what is meant by a *statistical connection* between data and give examples of how the discovery of such connections may be of assistance to managers, with comments on the dangers which may be involved.

6. Explain what is meant by *a.* significance of correlation, *b.* autocorrelation, *c.* covariance.

# 13

# Rates of Change

A more mathematical treatment of this subject will be found in chapter 21 where series are dealt with in detail.

## 13.1 SIMPLE INTEREST GROWTH

When we are dealing with time series a straight line graph is an indication of an equal absolute amount of growth for each time interval shown on the horizontal axis.

Such conditions may be referred to as *simple interest* growth since the graph of the amount of a principal invested at simple interest has this shape. The intercept on the y axis is the starting principal at the beginning of the time intervals, i.e. at $x=0$. The steepness of the rise is governed by the rate of interest.

A negative slope indicates that the value is reducing as time goes by. Probably the most usual illustration is afforded by straight line depreciation, where the same sum for depreciation is deducted each year.

A simple interest type of graph enables us readily to extrapolate the amount at any desired time simply by extending the line.

## 13.2 COMPOUND INTEREST GROWTH

In the case of *compound interest* it is the percentage or *ratio of growth* that remains constant throughout the graph. The amount at the start of each time interval is increased by the same percentage, so that the absolute amount of the increase becomes larger with the passage of time.

Table 13.1 shows the amounts of £100 at both simple and compound interest. Obviously the compound interest graph would lie above the simple interest curve, pulling away from it increasingly.

Table 13.1. *Simple and compound interest amounts of £100 at 5%*

| Year | Simple Interest | Compound Interest | Log. Comp. Int. |
|---|---|---|---|
| 0 | 100 | 100 | 2·0000 |
| 1 | 105 | 105·0 | 2·0212 |
| 2 | 110 | 110·3 | 2·0425 |
| 3 | 115 | 115·8 | 2·0637 |
| 4 | 120 | 121·6 | 2·0849 |
| 5 | 125 | 127·6 | 2·1059 |
| 6 | 130 | 134·0 | 2·1271 |
| 7 | 135 | 140·7 | 2·1483 |
| 8 | 140 | 147·7 | 2·1694 |
| 9 | 145 | 155·1 | 2·1904 |
| 10 | 150 | 162·9 | 2·2119 |

The last column of this table shows the logarithms of the compound interest amounts. It will be seen that these rise steadily, each being 0·0212 greater than the value before it. This difference is the logarithm of 1·05, the ratio of each value to the amount on which it is based.

The formula for compound interest takes the form

$$A=P\left(1+\frac{R}{100}\right)^N$$

where A is the amount of principal P invested for N years at R% interest. In the case of 5% the items are in geometric progression with a common ratio of $1+\frac{5}{100}=1{\cdot}05$.

The formula thus becomes $A=P(1{\cdot}05)^N$ and, taking logarithms, we have $\log A=\log P+N\log 1{\cdot}05$, which is a straight line connecting log A and N, having a constant term, logP, and a gradient, log1·05.

## 13.3 RATIO (OR LOGARITHMIC) SCALE GRAPHS

The property of logarithms to convert equal proportional rates of growth into a straight line graph can be made use of, either by plotting the logs of the ordinates on ordinary graph paper, or, more conveniently, by using *semi-logarithmic graph paper*, where the vertical scale is spaced in proportion to the logarithms of the numbers, so that the data can be plotted directly.

An example of the use of this type of graph is given in diagram 13.1, which illustrates the data of table 13.2.

Table 13.2. *Output (thousand tons) of firms A and B*

| Year | A | B |
|---|---|---|
| 1 | 3 | 43 |
| 2 | 4 | 44 |
| 3 | 5 | 45 |
| 4 | 6 | 46 |
| 5 | 3 | 23 |
| 6 | 9 | 69 |
| 7 | 12 | 92 |

Examination of the original data shows:

a. The outputs of both firms increase by the same amount, one unit, per year, up to year 4. On an ordinary graph these would appear as parallel lines. Over this period the proportional increase for firm B is small, while that for firm A totals 100%. This is reflected in the much more steeply rising curve for firm A on the logarithmic graph.

b. For the years 5, 6 and 7 both series experience the same proportional changes from one year to the next, a 50% fall, a 200% rise and a 33·3% rise. These equal proportional changes are reflected in the parallel lines on the logarithmic graphs for this period.

With logarithmic graphs it is the gradient which is vital to interpretation, not the vertical position of the graph.

The dotted lines of the diagram show how these graphs can be moved up or down so that they coincide at that point where we are interested in making a comparison of the growth. This is most easily done by tracing one of the lines and then moving into the required position relative to the other. Do not forget that it is the gradients of the graphs we are comparing. In the upper comparison the line for A has a steeper gradient on both sides of its intersection with the line for B, showing that its rate of growth immediately before and after the third year is greater. The equal gradients after the fourth year have already been mentioned, but are brought out more clearly with the graphs close together.

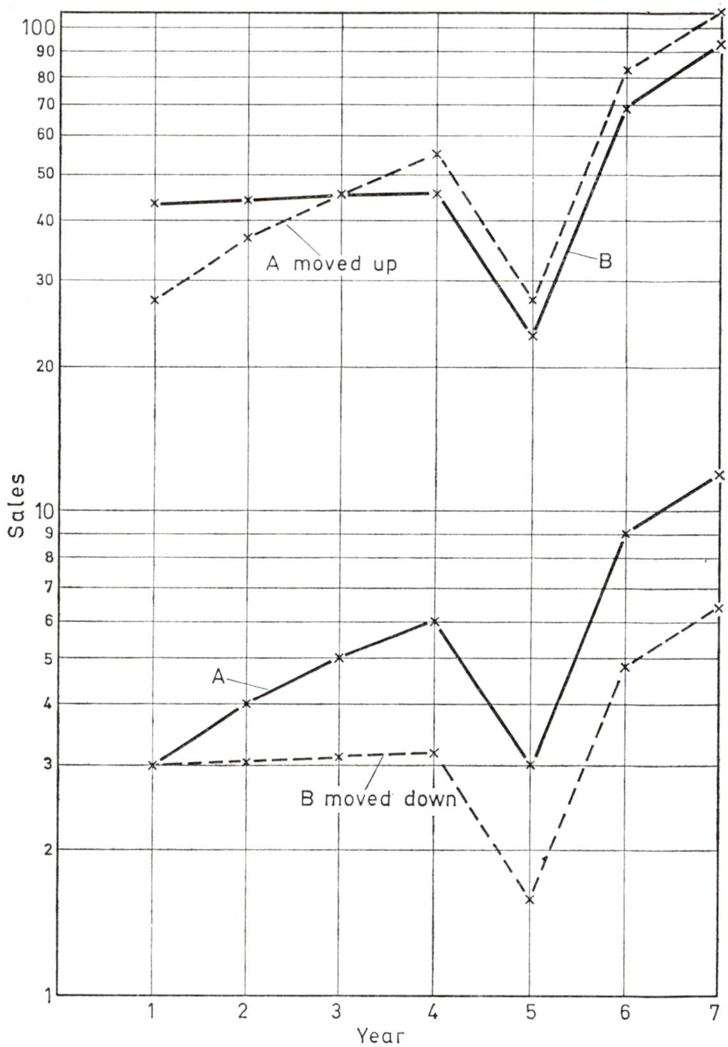


Diagram 13.1. *Use of semi-logarithmic graph paper* to compare sales of firms A and B

The percentage change between any two points on a ratio graph may be found by measuring the vertical distance between them, using a piece of paper to mark this off, and then measuring this along the logarithmic scale of the graph.

As an example, the distance between the values for years 5 and 7 for graph B comes to 4 log units, starting from 1. This shows a rise to 400%, equivalent to an increase of 300% on the 5th year figure.

It must be remembered that a 25% increase, say from 100 to 125, if reversed gives a 20% decrease, from 125 to 100. Thus the 4 log units, if taken to indicate the fall from year 7 to year 5, show this to be a fall to $\frac{1}{4}$ or a drop of 75%.

Semi-logarithmic graphs are frequently used to plot economic data and population statistics over the years since, in these cases, it is often the rate of change, rather than its actual amount, that is of interest.

If a straight line is obtained by this process we may fit a least squares regression line showing the relationship between the log of the data and the year. From this the rate of growth may be estimated. An example is given at the end of this chapter.

Any equation of the type $y=AB^x$ will result in a straight line if plotted on semi-logarithmic graph paper, since $\log y=\log A+x\log B$, and $\log A$ and $\log B$ are constants.

If the equation is of the type $y=Ax^B$ then $\log y=\log A+B\log x$, so that $\log y$ will give a straight line if plotted against *logx*. We may take the logs of both sets of data and plot on ordinary graph paper. Alternatively, we may plot on double-logarithmic graph paper, where both scales are in logarithmic form.

An important example of this type of curve is afforded by *Pareto's Law*, arising from studies of income distribution. It is a *higher the fewer* kind of curve. The simplified equation is $y=Ax^{-B}$, i.e. $y=\dfrac{A}{x^B}$, where y is the number of persons having an income of £x, and A and B are constants. If the values of y and x are plotted on double-logarithmic graph paper then the gradient is B.

## 13.4 NATURAL (EXPONENTIAL) GROWTH CURVES

Compound interest growth is an example of what scientists refer to as the natural law of growth, where the rate of increase of y is proportional to y itself.

The general form of the curve is $y=Ae^{Bx}$, where A and B are constants, and e is the *natural* number (see section 21.3 for a fuller explanation).

This natural growth curve is increasing all the time at a rate proportional to its value. In the case of compound interest increases are normally made at intervals of, say, a year or a month, and the curve is really a series of steps.

## 13.5 WORKED EXAMPLE

*Question*

The first two columns of table 13.3 shows the index of National Income for an imaginary country over five consecutive years. Calculate the yearly rate of growth, assuming this is of the compound interest type.

*Answer*

[If the growth is of the compound interest type the graph of the logarithms of the data plotted against the years should give a straight line. Inspection of the logs of the data, which have been entered in the third column, show they do increase by roughly equal amounts from year to year. The rest of the table gives the short-cut calculation of the least squares regression line (see section 11.2.2).]

Table 13.3. *Index of national income over 5 years*

| Year (x) | Index (y) | Log Index (Y) | Dev. from centre yr. (D) | $D^2$ | YD |
|---|---|---|---|---|---|
| 1 | 202 | 2·3054 | −2 | 4 | −4·6108 |
| 2 | 211 | 2·3243 | −1 | 1 | −2·3243 |
| 3 | 221 | 2·3444 | 0 | 0 | 0 |
| 4 | 231 | 2·3636 | 1 | 1 | 2·3636 |
| 5 | 242 | 2·3838 | 2 | 4 | 4·7676 |
| | | | 10 | | +0·1961 |

We require only the gradient of the line, which is $\frac{0{\cdot}1961}{10}=0{\cdot}01961$. The anti-log of this is 1·046, which is the ratio of each year's value to the previous year's. The compound rate of interest is thus 100(1·046−1)=4·6% per annum.

## 13.6 PRACTICE QUESTIONS

1. Explain the uses of logarithmic (or ratio-scale) graphs.
2. The following figures relate to the yearly exports of a certain commodity in thousands of tons for 10 successive years: 30, 34, 38, 42, 47, 53, 60, 70, 81, 93.
   *a.* Plot this data on an ordinary graph.
   *b.* Plot the logarithms of the quantities.
   *c.* Comment on the shapes of the two graphs.
3. Use the method of least squares to estimate the yearly rate of growth of the data qu. 2 for the first 7 years only.

# 14

# Time Series—Trends

## 14.1 ANALYSIS

Time series show how data vary over time. A typical example is shown in table 14.1, where quarterly sales of Product A are tabulated.

Table 14.1. *Product A—quarterly sales (£ thousands) over 21 years*

| Year | Quarter 1st | 2nd | 3rd | 4th | Yearly Total |
|---|---|---|---|---|---|
| 1 | 11 | 23 | 5 | 16 | 55 |
| 2 | 10 | 22 | 5 | 14 | 51 |
| 3 | 9 | 19 | 4 | 12 | 44 |
| 4 | 10 | 21 | 5 | 12 | 48 |
| 5 | 13 | 25 | 7 | 20 | 65 |
| 6 | 14 | 29 | 7 | 19 | 69 |
| 7 | 13 | 25 | 6 | 20 | 64 |
| 8 | 10 | 20 | 5 | 17 | 52 |
| 9 | 12 | 25 | 6 | 16 | 59 |
| 10 | 13 | 28 | 7 | 26 | 74 |
| 11 | 16 | 30 | 8 | 27 | 81 |
| 12 | 16 | 31 | 8 | 22 | 77 |
| 13 | 13 | 27 | 6 | 19 | 65 |
| 14 | 14 | 25 | 7 | 26 | 72 |
| 15 | 17 | 37 | 8 | 21 | 83 |
| 16 | 18 | 34 | 9 | 27 | 88 |
| 17 | 17 | 35 | 9 | 24 | 85 |
| 18 | 13 | 25 | 7 | 20 | 65 |
| 19 | 16 | 30 | 8 | 25 | 79 |
| 20 | 19 | 40 | 9 | 27 | 95 |
| 21 | 20 | 41 | 11 | 27 | 99 |

Such series can often be broken down into components which can be separately studied and may be of great use for economic analysis and business forecasting.

The most usual analysis is as follows:

a. Trend, which indicates any general tendency in the data which dominates other variations in the long run. The graph of the sales data (diagram 14.1) shows there is a clear upward trend in the data, but this does not always result in the second of any two adjacent years being the greater.

b. Cyclical variations, which tend to occur in a regular pattern over a period of years, sometimes causing the data to exceed and sometimes to fall short of the trend.

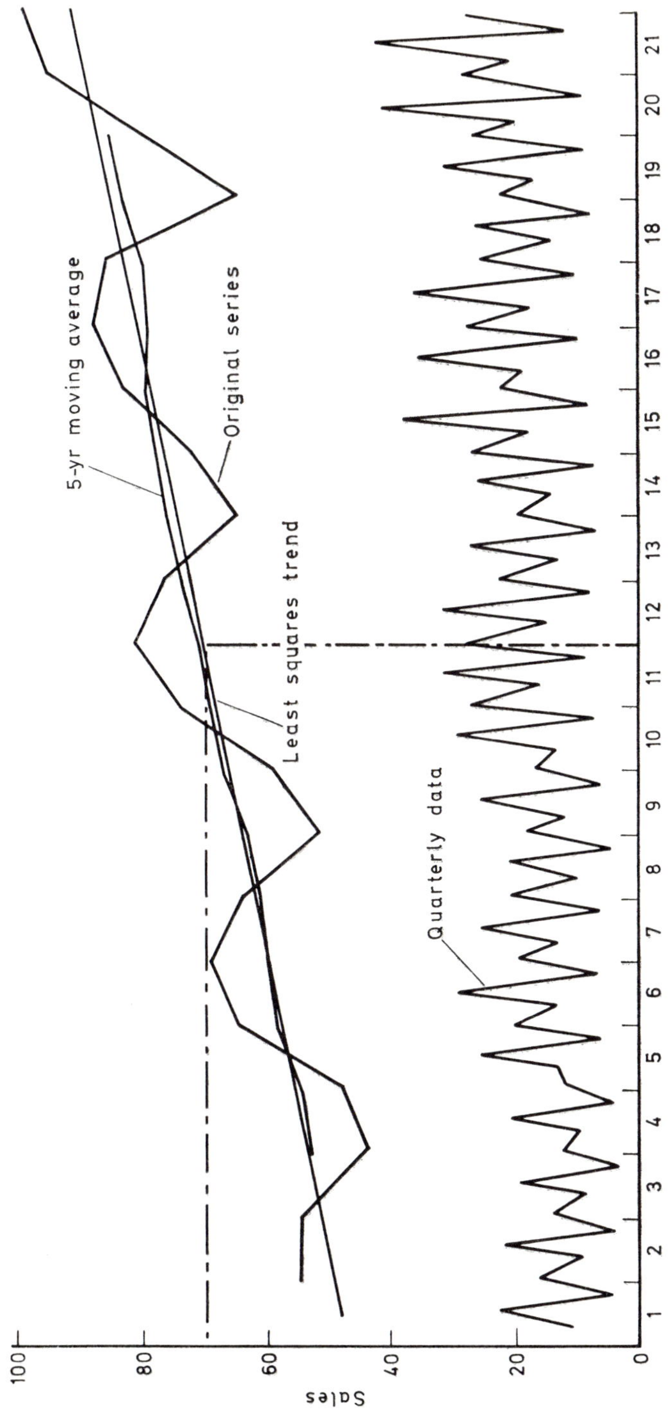


Diagram 14.1. *Time series and trends* relating to product A

c. Seasonal variations, which are similar to cyclical variations except that the pattern repeats every year, and we are concerned with, say, months or quarters. These can be very important for detailed budgets of businesses and for economic planning.

d. Residuals, which are the bits left over after an analysis on the lines just stated. They should be small relative to the variations that follow the lines of the analysis or the latter will be of little value.

e. Catastrophic variations, which are not capable of being treated statistically, except when aggregated, as in the case of fire risks for an insurance company. For the individual business a fire may cause its sales to drop to zero but it should be unpredictable.

Other approaches to the analysis problem are possible. Fourier analysis is an example, where the series is broken down into a number of sine curves.

After analysis we may be able to predict by pushing the trend forward and adjusting it for the appropriate variations. This can be dangerous unless there is reason to believe there have been no significant changes. *Extrapolation*, or reading values beyond the existing data, must always be handled with care.

Business forecasting usually implies the combination of two approaches:

a. *Time series analysis*, often called the *chartists approach* when stocks and shares are involved.

b. *Causal* or *market factors approach*, sometimes called the *fundamentalists* approach when applied to the stock markets, and involving estimates of the various factors that seem likely to correlate with the items being considered. For example sales will depend on population size and spending power, changes in taste, the strength of competition and the development of substitutes.

## 14.2 SIMPLE TREND ESTIMATION. SEMI-AVERAGES AND 3-POINT METHODS

The simplest form of trend is the *straight line*, and the simplest approach to estimation is to draw a line by guess by moving a ruler until its edge seems to represent the points. The yearly trend is then the gradient of this line (see chapter 11).

Some assistance in placing the line may be obtained by dividing the data into two parts, with an equal number of years in each. Find the average value of each half and plot it over the centre year of that half. Then join the two points.

If the number of years is odd the centre year's value may be added to each half, or omitted altogether. There is some advantage in having each half with an odd number of years since there will then be a centre year for each.

This method is sometimes called the method of *semi-averages*.

An alternative method is to divide the data into three equal sections on the basis of time. Plot the median values for the first and third sections over their centre time values and join these two points. Also find the median of all the data and plot it over the centre time of the whole data. The trend line is obtained by drawing a line parallel to the first line and one-third nearer to the overall median point.

If we wished to fit such a line for our specimen sales data we should divide the 21 years into 3 sections of 7 years each. For years 1 to 7 we should take the centre item *in size*, that is the 4th from the smallest, giving 55, which we should plot over year 4. For years 15 to 21 the median item is 85, which is plotted over the centre year of this group, the 18th year. These two points are then joined by a straight line. The median of all the items is the 11th in size and is the value 69 which occurs in the 6th year. This value is plotted over the centre year of all the data, which is the 11th year, and the former line is pulled one third of the way towards it. It is possible to carry this out using points drawn on a graph and just counting upwards to the various median values.

## 14.3 LEAST SQUARES AND ITS SIMPLIFIED APPLICATION

The most satisfactory way of fitting a straight line trend is by the method of least squares. This has been fully treated at 11.2.2.

We can use the basic normal equations approach to give a line of the form $y=A+Bx$, where A is the starting value of the data at year zero, x is the years beyond this, and B is the yearly trend increase, or decrease if B has a minus value.

Table 14.2. *Quick least squares trend calculation*

| Year | Deviation from Centre Year | Sales | | |
|---|---|---|---|---|
| x | d | y | $d^2$ | yd |
| 1 | −10 | 55 | 100 | −550 |
| 2 | −9 | 51 | 81 | −459 |
| 3 | −8 | 44 | 64 | −352 |
| 4 | −7 | 48 | 49 | −336 |
| 5 | −6 | 65 | 36 | −390 |
| 6 | −5 | 69 | 25 | −345 |
| 7 | −4 | 64 | 16 | −256 |
| 8 | −3 | 52 | 9 | −156 |
| 9 | −2 | 59 | 4 | −118 |
| 10 | −1 | 74 | 1 | −74 |
| 11 | 0 | 81 | 0 | 0 |
| 12 | 1 | 77 | 1 | 77 |
| 13 | 2 | 65 | 4 | 130 |
| 14 | 3 | 72 | 9 | 216 |
| 15 | 4 | 83 | 16 | 332 |
| 16 | 5 | 88 | 25 | 440 |
| 17 | 6 | 85 | 36 | 510 |
| 18 | 7 | 65 | 49 | 455 |
| 19 | 8 | 79 | 64 | 632 |
| 20 | 9 | 95 | 81 | 855 |
| 21 | 10 | 99 | 100 | 990 |
| Totals | | 1470 | 770 | 1601 |

Since time series generally have the dates equally spaced we can often use the quick method, taking deviations from the centre year.

This is shown in table 14.2. The trend is $\frac{\Sigma yd}{\Sigma d^2}=\frac{1601}{770}=2{\cdot}079$. The average of the years is 11 and of the data $\frac{1470}{21}=70$.

The trend equation is thus

$$y-70=2\cdot079(x-11)$$

$$y=47\cdot13+2\cdot079x$$

If we wish to estimate the trend value for, say, the 26th year of this series, we substitute 26 for x and find that the value of y is 101·2.

Sometimes it may be convenient to leave the series based on the centre year, when the equation is simply

$$y=70+2\cdot079d$$

in which d is the time deviation from the centre year.

The case where there are an even number of years can be treated by a slight modification of this method, of which an example is given at 14.8.1.

## 14.4 MOVING TOTALS AND AVERAGES

A straight line may not always represent the trend of data. The *moving average* provides a line which snakes through the points, turning to take account of each additional item as it discards the equivalent earlier one.

The basic idea is to eliminate cyclical or seasonal effects by including the years for a complete cycle (or the months or seasons for a whole year) in a total, which is then averaged. The total is progressively updated, always including a complete cycle (or year).

Table 14.3. *Calculation of 5-year moving average*

| Year | Sales | 5-Year Total | Trend |
|---|---|---|---|
| 1 | 55 | | |
| 2 | 51 | | |
| 3 | 44 | | 52·6 |
| 4 | 48 | | 55·4 |
| 5 | 65 | 263 | 58·0 |
| 6 | 69 | 277 | 59·6 |
| 7 | 64 | 290 | 61·8 |
| 8 | 52 | 298 | 63·6 |
| 9 | 59 | 309 | 66·0 |
| 10 | 74 | 318 | 68·6 |
| 11 | 81 | 330 | 71·2 |
| 12 | 77 | 343 | 73·8 |
| 13 | 65 | 356 | 75·6 |
| 14 | 72 | 369 | 77·0 |
| 15 | 83 | 378 | 78·6 |
| 16 | 88 | 385 | 78·6 |
| 17 | 85 | 393 | 80·0 |
| 18 | 65 | 393 | 82·4 |
| 19 | 79 | 400 | 84·6 |
| 20 | 95 | 412 | |
| 21 | 99 | 423 | |

With cyclical data we must first estimate the length of the cycle, i.e. the number of years from one peak to the next. In our example this period is 5 years. We then add the first 5 years, giving 263 (table 14.3). The next step is to discard the first year and add on the next year (6th) giving a total of 277. This is most easily done by adding the difference between 69 and 55, i.e. 14 to the existing value of 263 to give 277. This is repeated until the end of the data. These are 5 year totals.

If, instead of years, we had been working in terms of months or seasons, then these totals would have been *moving annual totals*, giving a full year's results up to the end of December, January, February etc., each including 12 months or 4 quarters.

The 5 year totals are each divided by the number of years included to give average values, which are shown in the *trend* column opposite the centres of the cycles to which they relate.

The method of least squares and the moving average are really special cases of the same approach. We know that the least squares line passes through the average of the data and its centre year. With the moving average we calculate the average at stages throughout the data and plot over the centre year of each stage. This has the effect of connecting the centroids of a number of least squares straight lines, each covering a cycle.

A disadvantage of the moving average method is that it does not provide trend figures for the first and last few periods of the data.

Where, as in this example, there is an odd number of years the centre of each cycle occurs at an actual year. If the number of years is even then the centre is between two actual values, e.g. with a 6-year cycle it is at year $3\frac{1}{2}$. This may be inconvenient for further working. The averages may be centred on actual years by taking two moving averages at a time, averaging them, and placing the result opposite the 4th year, then moving on to take the second and third averages and placing their average against the 5th year, and so on.

This can be important with seasonal variations, where even numbers commonly occur, e.g. 4 seasons or 12 months. The 13 month year has advantages from this point of view, as well as always containing the same number of weeks in each month.

If a seasonal or cyclical index is available (see chapter 15) this may be applied to the crude data to remove the seasonal or cyclical effect, leaving the trend. The crude data are divided by the index appropriate to the particular month or year, if the index is in the form of a proportion (and multiplied by 100 if it is in the form of a percentage). Where the index is in the form of a actual amount then this is deducted from the crude data (if it is +) or added to it (if it is −).

This process is sometimes called *deflating* the series.

## 14.5 EXPONENTIAL SMOOTHING. FILTERS

This is really a development from the moving average. The data included in each average is weighted so that the more recent data has a greater effect than the more remote. The weights are each a constant proportion of the weights for the following year and it is arranged that they total unity to simplify calculation.

The proportion used is at the discretion of the statistician. If, for example, we take the proportion as 0·6 this implies that the weights are proportional, *going back through the years*, to 1, 0·6, $(0{\cdot}6)^2$, $(0{\cdot}6)^3$, $(0{\cdot}6)^4$ etc. The fact that we are dealing with powers of numbers is the reason for the name *exponential*. It will be seen that these weights are in geometric progression with a common ratio of 0·6 (see section 21.3.2). The total of such a series to infinity is

$\frac{1}{1-0{\cdot}6}$, so that, if we multiply each by $1-0{\cdot}6=0{\cdot}4$, the weights will total unity.

This gives weights of 0·4, 0·24, 0·14, 0·09 and 0·05, if five years are taken. The sum of these weights is actually 0·92 because the series has not enough terms to make the total unity.

If we apply these weights, smallest first, to the first five years of the sales data, then the weighted average is regarded as a trend estimate for the *fifth* year.

This is shown in table 14.4. The estimate for year 5 is then $\frac{51{\cdot}02}{0{\cdot}92}=55{\cdot}45$.

Table 14.4. *Exponential smoothing of sales data*

| Year | Sales | Weight | Sales × Weight |
|---|---|---|---|
| 1 | 55 | 0·05 | 2·75 |
| 2 | 51 | 0·09 | 4·59 |
| 3 | 44 | 0·14 | 6·16 |
| 4 | 48 | 0·24 | 11·52 |
| 5 | 65 | 0·04 | 26·00 |
| | | 0·92 | 51·02 |
| 6 | 69 | | |
| 7 | 64 etc | | |

Having obtained this estimate it can be updated very easily, when the next year's figure is available. We multiply the new item by 0·4 and add to it the old trend figure multiplied by 0·6. This gives $0{\cdot}4\times69+0{\cdot}6\times55{\cdot}45=60{\cdot}87$. The trend for year 7 is similarly $0{\cdot}4\times64+0{\cdot}6\times60{\cdot}87=62{\cdot}12$, and so on. This arises because of the constant proportions between the weights.

It is possible to deal with this sort of calculation very easily on a computer, using minimum storage, since only two values are used. This makes it very suitable for forecasting where, say, predictions have to be made about large numbers of stock items.

Because of its differential weighting this method is not suited to smoothing data where there is a pronounced seasonal or cyclical element. This has to be eliminated to start with from the crude data.

It should also be noted that the method does not of itself provide a forecast, or an indication of the trend. It is intended to provide a stable basis for the year concerned, to which trend adjustments etc. can be made.

A simplified smoothing technique avoids the initial calculation based on several periods. The first year is chosen as a base. The second year is smoothed by multiplying the first year value by $\alpha$ and adding to it the second year value multiplied by $1-\alpha$. The third year is similarly a weighted average of the second year smoothed value and the third year actual value.

Let us take the data of table 14.1 and suppose that $\alpha$ is to be 0·4. The smoothed values are as follows:

| | |
|---|---|
| 1st year | 55 (actual) |
| 2nd year | $0{\cdot}4\times55+0{\cdot}6\times51=52{\cdot}6$ |
| 3rd year | $0{\cdot}4\times52{\cdot}6+0{\cdot}6\times44=47{\cdot}4$ etc. |

An alternative starting value is sometimes taken as the ordinary mean of a number of periods.

The technique illustrated is sometimes termed applying a *smoothing filter* to the data.

The higher the value of $\alpha$ the greater is the weight given to the past and the greater the smoothing effect. For short periods of time $\alpha$ is usually taken at a higher value than for long periods. With monthly data we might set $\alpha$ at 0·7 or 0·8, while for annual data 0·4 or 0·3 might be most appropriate.

## 14.6 SHORT-TERM FORECASTS. HOLT AND BOX-JENKINS

*Holt's* method of forecasting estimates the sales trend figure for the next period as the weighted sum of the actual sales in the latest (current) period and the forecast of these sales made in the previous period.

The weighting incorporated is a value $\alpha$ (say 0·1, 0·2, or 0·3 as experience shows to be appropriate).

Forecast for the next period=actual latest sales $\times$ $\alpha$+forecast latest sales $\times$ $(1-\alpha)$.

Rearrangement of the terms shows this to be the same as:

Forecast for the next period=forecast of latest sales+$\alpha$ (actual latest sales – forecast latest sales).

The term in the bracket is the error in the previous forecast (with its sign reversed). We are thus taking the last forecast and adjusting it by a predetermined fraction of this error.

*Box-Jenkins*' predictor offers a similar but more complicated approach, taking account of the cumulative error sum.

It must be emphasised that models such as Holt's are suited to short-term forecasting where stable values with little seasonal variation are concerned.

Suppose that we are dealing with such a commodity and that last month the sales were forecast as 300 units whereas in fact they turned out to be 280 units. The difference is $280-300=-20$ units. If we assume $\alpha$ to be 0·2 then the estimate for the next month is $300-0{\cdot}2\times20=296$ units. Alternatively, applying the first version of the formula, we have $280\times0{\cdot}2+300\,(1-0{\cdot}2)=296$.

## 14.7 SOME SPECIAL TRENDS

Sometimes a straight line may be unsatisfactory to express a trend.

A *parabola*, of the form $y=A+Bx+Cx^2$ may be suitable. This may be fitted by a more complicated version of the method of least squares, involving three normal equations, in place of the two used in section 11.2.

An *exponential*, or power curve, of the form $y=AB^x$, may suit conditions of growth or decay of the compound interest type. We may test for the suitability of such a curve by plotting logy against x (or using semi-logarithmic graph paper), when a straight line will indicate that this curve is appropriate. Since $\log y=\log A+x\log B$ we may apply the method of least squares to find the relationship between logy and x, and so find the constants.

The general form of the exponential curve is $y=e^{(A+Bx)}$. There is a *modified exponential* curve of the type $y=A+BC^x$.

The *Gompertz* curve increases at a diminishing rate, until it reaches a maximum value, and may be suited to forecasting the development of a new product to saturation point. The curve takes the form $y=AB^{c^x}$ or $\log y=\log A+C^x\log B$.

The *logistic* curve, of the form $y=\frac{1}{A+BC^x}$, is also suitable to represent forms of economic growth associated with expansion to a saturation level.

### 14.8.1 WORKED EXAMPLE

*Question*

Table 14.5 (first two columns) shows the quantity of an item *y* over the eight years from 1960 to 1967. Use the method of least squares to estimate the values for the next four years, assuming that the trend is a straight line.

**Table 14.5**

| Year | Data (y) | Years from Start | Deviation (D) from $4\frac{1}{2}$ in $\frac{1}{2}$'s | $D^2$ | yD |
|---|---|---|---|---|---|
| 1960 | 36·5 | 1 | −7 | 49 | −255·5 |
| 1961 | 38·7 | 2 | −5 | 25 | −193·5 |
| 1962 | 43·6 | 3 | −3 | 9 | −130·8 |
| 1963 | 46·9 | 4 | −1 | 1 | −46·9 |
| 1964 | 50·1 | 5 | 1 | 1 | 50·1 |
| 1965 | 53·4 | 6 | 3 | 9 | 160·2 |
| 1966 | 54·6 | 7 | 5 | 25 | 273·0 |
| 1967 | 58·3 | 8 | 7 | 49 | 408·1 |
| | 382·1 | | | 168 | +264·7 |

*Answer*

[In this case there are an even number of evenly spaced x values (years) so we may use the short cut method. The actual dates have been converted to years from the start. The centre value is $4\frac{1}{2}$. Because of this we work in deviation units of $\frac{1}{2}$ a year, giving the column D. These values are squared and summed, and also multiplied by the data, which column is also summed.]

$$\text{Average value of data}=\frac{382{\cdot}1}{8}=47{\cdot}76.$$

$$\text{Half yearly increase}=\frac{264{\cdot}7}{168}=1{\cdot}58.$$

Yearly increase is twice this i.e. 3·16.

The equation for the value of y in terms of years from the start is given by

$$y-47{\cdot}76=3{\cdot}16\,(x-4{\cdot}5)$$

from which, $y=33{\cdot}54+3{\cdot}16x$, where x is the year.

The trend values for the years 9, 10, 11 and 12 are found by substituting these values to establish y, giving the following:

| | |
|---|---|
| 1968 | 62·0 |
| 1969 | 65·1 |
| 1970 | 68·3 |
| 1971 | 71·5 |

[Another way of arriving at these results is to consider that the average value of 47·76 refers to the year ending half way through year 5. The value for year 5 is thus half of this plus another half increased by the half-yearly increment, giving $47{\cdot}76+1{\cdot}58=49{\cdot}34$. This is the value for year 5, or 1964, so that the 1968 estimate is $49{\cdot}34+4\times3{\cdot}16=61{\cdot}98$, etc.]

## 14.8.2 WORKED EXAMPLE

*Question*

Use the quarterly figures for the first five years of table 14.1 and remove the seasonal variation by means of a moving average.

Table 14.6

| Year | Qtr. | Sales | M.A.T. | M.A.T. ÷4 | Trend |
|---|---|---|---|---|---|
| 1 | 1 | 11 | | | |
| | 2 | 23 | | | |
| | 3 | 5 | ........... | ........... | 13·625 |
| | 4 | 16 | 55 | 13·75 | 13·375 |
| 2 | 1 | 10 | 54 | 13·50 | 13·250 |
| | 2 | 22 | 53 | 13·25 | 13·000 |
| | 3 | 5 | 53 | 13·25 | 12·625 |
| | 4 | 14 | 51 | 12·75 | 12·125 |
| 3 | 1 | 9 | 50 | 12·50 | 11·625 |
| | 2 | 19 | 47 | 11·75 | 11·250 |
| | 3 | 4 | 46 | 11·50 | 11·125 |
| | 4 | 12 | 44 | 11·00 | 11·500 |
| 4 | 1 | 10 | 45 | 11·25 | 11·875 |
| | 2 | 21 | 47 | 11·75 | 12·000 |
| | 3 | 5 | 48 | 12·00 | 12·375 |
| | 4 | 12 | 48 | 12·00 | 13·250 |
| 5 | 1 | 13 | 51 | 12·75 | 14·000 |
| | 2 | 25 | 55 | 13·75 | 15·250 |
| | 3 | 7 | 57 | 14·25 | |
| | 4 | 20 | 65 | 16·25 | |

*Answer*

[The calculation is shown in table 14.6. It will be seen that a moving annual total has first been formed by taking four quarters together at a time. These figures have then been averaged. Since the centres would be at 2½, 3½ etc. these averages have themselves been averaged, two at a time, and the results placed opposite the 3rd, 4th etc. quarters. It is tempting to use five quarters at a time to provide a centre quarter but this is appropriate only when pronounced seasonal fluctuations are absent.]

## 14.9 PRACTICE QUESTIONS

1. *a.* Determine the least squares trend line to fit the following data of gross coal consumption for the years 1957—1967 (millions of tons).

   *b.* From this estimate the consumption for the next 3 years. 212·9, 202·4, 189·4, 196·7, 191·8, 191·2, 194·0, 187·2, 184·6, 174·7, 163·9 (Ministry of Power).

2. The following figures relate to the mid-year estimates of U.K. population for each year from 1951 to 1967 (millions): 50·3, 50·4, 50·6, 50·8, 50·9, 51·2, 51·4, 51·7, 52·0, 52·4, 52·8, 53·3, 53·7, 54·1, 54·4, 54·7, 55·1 (Registrars General).

   *a.* Plot these on a graph so as to emphasise the trend.

   *b.* Fit a straight line by the method of least squares to the last 9 years.

*c.* Suggest any other form of trend line which might be suitable for the whole of the data.

3. Use the three-point method to fit a trend line to the first 15 years of the data of qu. 2.

4. Use a moving average to smooth the data of qu. 4 of 15.5.

5. Calculate the moving annual totals for the data of qu. 1 of 15.5.

6. Smooth the data of qu. 1 by means of a suitable filter.

7. Smooth the data of qu. 3 of 15.5 using a moving average.

8. Compare forecasting based on fitting a trend line to data with techniques derived from exponential smoothing.

# 15

# Time Series—Variations

### 15.1.1 CYCLICAL AND SEASONAL VARIATIONS

The nature of these variations has been explained at 14.1. They result in variations from the trend and follow a fairly regular up and down pattern over the *cycle time* involved. Because the pattern repeats itself throughout the data it is usually possible to indicate its features in the form of a cyclical or seasonal index. Broadly the same techniques are involved in calculation, whether it is a case of an index showing how the years of a cycle vary, or of one representing the seasons or months of a year.

Two decisions have to be made:

a. Is the index to be calculated *directly* from the crude data, or shall we *first estimate the trend value* corresponding to each actual value and remove the effect of this by expressing the actual as a percentage of the trend? The latter technique can be laborious but it gives a superior result. If the trend is not first removed a rising trend, for example, will tend to overstate the later periods of the index at the expense of the former.
b. Is the index to be in the form of a *ratio*, where 1 or 100 represents the average situation, or is it to be an *absolute* amount to be added to or subtracted from a trend figure? The proportional index has the merit that it is related to the size of the item to which it is applied.

### 15.1.2 DIRECT ESTIMATION

The calculation of an index for each of the five years of the cycle in the sales data already used is shown in table 15.1.

Table 15.1. *Calculation of cyclical index from crude sales data*

| | Year of Cycle | | | | | |
|---|---|---|---|---|---|---|
| Cycle | 1 | 2 | 3 | 4 | 5 | Total |
| 1 | 55 | 51 | 44 | 48 | 65 | 263 |
| 2 | 69 | 64 | 52 | 59 | 74 | 318 |
| 3 | 81 | 77 | 65 | 72 | 83 | 378 |
| 4 | 88 | 85 | 65 | 79 | 95 | 412 |
| 5 | 99 | — | — | — | — | 99 |
| Totals | 392 | 277 | 226 | 158 | 317 | 1470 |
| Averages | 78·4 | 69·3 | 56·5 | 64·5 | 79·3 | 70·0 |
| Indices | 112·0 | 99·0 | 80·7 | 92·1 | 113·3 | (=total 497·1) |
| Adjusted Indices | 112·6 | 99·6 | 81·3 | 92·7 | 113·9 | |

The original items are copied into this table so that each column contains a similar year of each cycle. The columns are totalled and averaged by dividing by the number of items in each. These averages are expressed as percentages of the overall average $\left(\frac{1470}{21}=70 \text{ in this case}\right)$, to give the indices 112, 99 etc.

The total of these indices should be N×100, where N is the number of indices we are calculating. If it is not then the difference is divided evenly between each index so as to give the correct total. In the case shown the error is 500—497·1=2·9. This must be divided by 5 to give the amount to be added to each index, viz. $\frac{2\cdot9}{5}=0\cdot6$.

The final result is shown in the last row of the table. Actually the total would now be 500·1, but the last, and largest index, has been called 113·8 instead of 113·9, since we do not wish to go beyond one place of decimals.

Table 15.2. *Calculation of seasonal indices from crude sales data*

| | Season 1 | 2 | 3 | 4 | Total |
|---|---|---|---|---|---|
| Totals | 294 | 592 | 147 | 437 | 1470 |
| Averages | 14.0 | 28·2 | 7·0 | 20·8 | 17·5 |
| Index | 80·0 | 161·1 | 40·0 | 118·9 | (=total 400·0) |

Table 15.2 shows a similar calculation to provide a seasonal index from the crude data. The quarterly columns of table 14.1, which are already appropriately grouped for seasons, have been added to give the totals of the first row. Each of these has been divided by 21, the number of items included in it, and the resulting averages have been expressed as percentages of the average quarterly figure overall $\left(\frac{1470}{84}=17\cdot5\right)$ to give the required indices. Since these total 400·0 there is no further adjustment needed.

### 15.1.3 REMOVAL OF TREND BEFORE ESTIMATION

If the trend is to be eliminated from the crude data before the cyclical or seasonal indices are calculated this can be done as in table 15.3.

The least squares trend has been chosen and the trend value calculated for each year. Two approaches are then used, which would normally be regarded as alternatives. The data may be either shown as a percentage of the trend (which is more difficult from the point of view of calculation, but perhaps more useful) or as a deviation from it. A third possibility, not shown, is to express the deviations as + and − percentage variations of the trend. This is the same as subtracting 100 from each of the percentage figures in the fourth column of the table.

Calculation of the cyclical index can be based on the figures of either column 4 or 5. These figures will replace the crude data in table 15.1, the calculation being on precisely similar lines.

Readers may care to work through this calculation for the percentaged figures. The indices, in order, came out as 112·9, 103·9, 82·7, 91·1 and 109·4. These should be compared with the similar indices based on the crude data, shown in table 15.1.

Table 15.3. *Removal of least squares trend from sales data*

| Year | Data | Trend | Data as % of Trend | Variation of Data from Trend |
|---|---|---|---|---|
| 1 | 55 | 49·2 | 111·8 | 5·8 |
| 2 | 51 | 51·3 | 99·4 | −0·3 |
| 3 | 44 | 53·4 | 82·5 | −9·4 |
| 4 | 48 | 55·5 | 86·6 | −7·5 |
| 5 | 65 | 57·5 | 113·0 | 7·5 |
| 6 | 69 | 59·6 | 115·8 | 9·4 |
| 7 | 64 | 61·7 | 103·8 | 2·3 |
| 8 | 52 | 63·8 | 81·6 | −11·8 |
| 9 | 59 | 65·8 | 89·6 | −6·8 |
| 10 | 74 | 67·9 | 109·0 | 6·1 |
| 11 | 81 | 70·0 | 115·7 | 11·0 |
| 12 | 77 | 72·1 | 106·8 | 4·9 |
| 13 | 65 | 74·2 | 87·7 | −9·2 |
| 14 | 72 | 76·2 | 94·4 | −4·2 |
| 15 | 83 | 78·3 | 106·0 | 4·7 |
| 16 | 88 | 80·4 | 109·5 | 7·6 |
| 17 | 85 | 82·5 | 103·1 | 2·5 |
| 18 | 65 | 84·6 | 76·9 | −19·6 |
| 19 | 79 | 86·6 | 91·2 | −7·6 |
| 20 | 95 | 88·7 | 107·1 | 6·3 |
| 21 | 99 | 90·8 | 109·0 | 8·2 |

The calculation of the cyclical indices based on deviations of the data from the trend is shown in table 15.4. The deviations are grouped in the usual way, totalled and averaged. Since, over the cycle, such deviations should cancel out, their total should be zero. In this case it is −2, so we must add $\frac{2{\cdot}0}{5}=0{\cdot}4$ to each item, giving the indices as shown.

Table 15.4. *Calculation of cyclical variations based on absolute variations from the trend*

| | Year of Cycle | | | | | |
|---|---|---|---|---|---|---|
| Cycle | 1 | 2 | 3 | 4 | 5 | |
| 1 | 5·8 | −0·3 | −9·4 | −7·5 | 7·5 | |
| 2 | 9·4 | 2·3 | −11·8 | −6·8 | 6·1 | |
| 3 | 11·0 | 4·9 | −9·2 | −4·2 | 4·7 | |
| 4 | 7·6 | 2·5 | −19·6 | −7·6 | 6·3 | |
| 5 | 8·2 | — | — | — | — | |
| Totals | 42·0 | 9·4 | −50·0 | −26·1 | 24·6 | |
| Averages | 8·4 | 2·4 | −12·5 | −6·5 | 6·2 | (=total −2·0) |
| Indices | 8·8 | 2·8 | −12·1 | −6·1 | 6·6 | |

These indices would be added to, or deducted from the trend, depending on the sign, to give an estimate of the yearly sales, in a forecast. Alternatively, crude data may have the cyclical variations removed by subtracting plus values and adding minus values.

The trend value at the start of the series is about 50 and rises to about 90 for the final year. With this type of absolute index we should, nevertheless, make the same adjustment to each figure, which does not seem as reasonable as adjusting each by about 13%, as would be the case with the proportional index for the first year of the cycle.

The absolute index can be converted to a proportional one by dividing it by the average of the data (70) and multiplying by 100, if it is required in the percentage form. The result is an index of the percentage up or down for each year. If the average of 100 is added to each of these we arrive at the indices previously calculated from the percentaged figures.

## 15.2 NORMALISING (OR DEFLATING) A TIME SERIES. RESIDUALS

If a cyclical or seasonal index is available it may be applied to the crude data to indicate the trend.

Table 15.5. *Removing cyclical effects from a time series*

| Year | Data | Index | Adjusted Data | Least Squares Trend | Col. 4 −Col. 5 |
|---|---|---|---|---|---|
| 1 | 55 | 112·9 | 48·7 | 49·2 | −0·5 |
| 2 | 51 | 103·9 | 49·1 | 51·3 | −2·2 |
| 3 | 44 | 82·7 | 53·2 | 53·4 | −0·2 |
| 4 | 48 | 91·1 | 52·7 | 55·5 | −2·8 |
| 5 | 65 | 109·4 | 59·4 | 57·5 | 1·9 |
| 6 | 69 | 112·9 | 61·1 | 59·6 | 1·5 |
| etc. | | | | | |

This has been done to part of the sales data in table 15.5. Each item has been divided by the index appropriate to the particular year and then multiplied by 100. The least squares trend values are shown for comparison. The difference between the two estimates is an indication of the *residuals* effects, i.e. those that do not conform exactly to the assumptions on which the analysis is based.

Alternatively, we may start with the least squares trend values and adjust each by multiplying by the appropriate index and dividing by 100, to give an estimate of what the data should be. This may be compared with the actual data, and the residual difference found in each case. The reader should try this calculation.

## 15.3 INDICES IN FORECASTING

Indices may be applied in forecasting in two ways.

First, we may have extrapolated on the basis of an established trend line to obtain an estimate of the normal value for some future year. This forecast may be made more realistic by multiplying it by the appropriate cyclical index (dividing by 100 if this is in percentage form). At section 14.3, for example, we estimated the sales for the 26th year to be 101·2. The 26th year is the first year of the appropriate 5-year cycle. A better estimate of the sales for the 26th year is thus $101{\cdot}2 \times \frac{112{\cdot}9}{100} = 114{\cdot}3$, using the index shown at 15.1.3.

Alternatively, having established an estimate for a future period, we may wish to break it down into realistic estimates of its components. The figure of 114·3 is for the 26th year, but how should it be divided between the seasons? If we divide it by the number of seasons, to arrive at the average seasonal estimate, we may then apply the appropriate seasonal indices, in the same way as we have just used the cyclical ones.

The average seasonal value is $\frac{114 \cdot 3}{4} = 28 \cdot 6$. In the first season of this year, for which the index is 80·0 (table 15.2), the expected sales would be $28 \cdot 6 \times \frac{80 \cdot 0}{100} = 22 \cdot 9$. The values for the other seasons can be similarly calculated, and the total of the seasonal estimates should be checked to see that it agrees with the year's total.

## 15.4 WORKED EXAMPLE

*Question*

A firm engaged in installation work estimates that, in the following year, it can sign up 108,000 new customers to have this work done. The work is highly seasonal, and the management wish to estimate the likely monthly load to help with planning. Past experience shows that the monthly indices appropriate to this are, starting with January, 40, 45, 75, 90, 100, 125, 150, 170, 140, 110, 95, 60.

*a.* Estimate the monthly loading.

*b.* Produce a cumulative target throughout the year.

*c.* Suggest, but do not draw, two forms of diagram that might be used in controlling this situation during the year in question.

*Answer*

*a.* and *b.* The calculation is shown in table 15.6.

The average monthly sales $= \frac{108{,}000}{12} = 9{,}000$

The monthly loadings are thus $9{,}000 \times \frac{40}{100} = 3{,}600$,

$9{,}000 \times \frac{45}{100} = 4{,}500$ etc.

Table 15.6. *Estimates of monthly loading and cumulative targets*

| Month | Index | Monthly Loading | Cumulative Target |
|---|---|---|---|
| 1 | 40 | 3600 | 3600 |
| 2 | 45 | 4050 | 7650 |
| 3 | 75 | 6750 | 14400 |
| 4 | 90 | 8100 | 22500 |
| 5 | 100 | 9000 | 31500 |
| 6 | 125 | 11250 | 42750 |
| 7 | 150 | 13500 | 56250 |
| 8 | 170 | 15300 | 71550 |
| 9 | 140 | 12600 | 84150 |
| 10 | 110 | 9900 | 94050 |
| 11 | 95 | 8550 | 102600 |
| 12 | 60 | 5400 | 108000 |
| | 1200 | 108000 | |

The cumulative targets are found by adding each month's value to the total of the previous months.

*c.* A Gantt chart might be used to show monthly and cumulative targets and actual results.
A Z-chart could also be used.
[Details of these are given in chapter 14, using the figures of this question.]

## 15.5 PRACTICE QUESTIONS

1. The following table shows the home deliveries of phosphate chemical fertilisers (thousand tons) from January to December for the three years 1966 to 1968

| Year | Month | | | | | |
|---|---|---|---|---|---|---|
| | 1 | 2 | 3 | 4 | 5 | 6 |
| 1966 | 35·1 | 39·5 | 62·3 | 34·0 | 31·1 | 46·6 |
| 1967 | 39·5 | 43·3 | 68·0 | 46·5 | 23·3 | 46·9 |
| 1968 | 33·7 | 38·9 | 58·5 | 42·5 | 22·7 | 50·5 |
| | 7 | 8 | 9 | 10 | 11 | 12 |
| 1966 | 40·0 | 28·0 | 29·9 | 25·6 | 28·4 | 26·6 |
| 1967 | 44·6 | 29·3 | 34·0 | 33·2 | 26·8 | 31·7 |
| 1968 | 50·5 | 29·6 | 28·3 | 30·0 | 29·7 | — |

(Board of Trade)

Using the crude data, without trend adjustment, calculate the monthly seasonal indices.

2. Group the data of qu. 1 into quarters and calculate a quarterly seasonal index.

3. The quarterly totals (million pairs) of manufacturers' sales of footwear are as follows:

| Quarter | 1965 | 1966 | 1967 | 1968 |
|---|---|---|---|---|
| 1 | 47·7 | 51·3 | 48·2 | 50·7 |
| 2 | 47·2 | 46·1 | 45·3 | 46·1 |
| 3 | 50·6 | 49·1 | 47·5 | 51·1 |
| 4 | 52·2 | 47·3 | 49·1 | — |

(Board of Trade)

*a.* Prepare a seasonal index from the data.

*b.* How could this be used practically?

4. The cyclical indices, covering a 5-year cycle, for certain data are as follows: 75, 80, 125, 150, 70. Actual figures over a period of 14 years, starting with the first year of the cycle are: 150, 170, 270, 330, 165, 190, 205, 340, 425, 200, 220, 260, 380, 480.

*a.* Use the indices to normalise the series.

*b.* From the crude data calculate a new cyclical index.

5. The estimated sales of an article for a particular year are £28,000, based on a least squares trend calculation. The cyclical index for this year is 90. What is a reasonable estimate of sales for March assuming that the monthly index for March is 155?

# 16

# Diagrams

## 16.1 GENERAL CONSIDERATIONS

Diagrams and graphs may serve two purposes:

a. Sometimes they are *aids to calculation*, as when a graph is used to read off the y values corresponding to chosen values of x, or when two graphs are plotted and their point of intersection gives the solution of the equations. In cases of this kind the main consideration is the accuracy of the result. Use of a large scale generally helps and we may well just graph that part of the curve(s) relevant to the solution of the problem. We are not especially concerned with the impression made on others, though accuracy is essential and neatness contributes to this.

b. Often the diagrams serve the quite different purpose of *conveying information*, perhaps to those not expert in their subject matter. A good diagram will arouse interest and then quickly convey the essential facts. Detail will be omitted or relegated to the background. It is essential that, in presenting this simplified picture, a wrong impression is not likely to arise.

It is with the second type of diagram that this chapter is concerned.

## 16.2 SOME RULES

It is difficult to formulate precise rules for drawing diagrams since much depends on the flair and artistic appreciation of the drawer. The following points should be considered, however:

a. People find it much easier to make comparisons between the lengths of straight lines than of more complex shapes. A bar chart is less likely to mislead than a pie diagram or an attempt at a three dimensional diagram. When a comparison of areas is involved these are in proportion to the square of any dimension, and when volumes are being compared these are related to the cube of any dimension. Experiments indicate that if two ships are drawn, the sizes of which are supposed to represent, say, the volumes of exports at two dates, then the number of different interpretations of viewers, and their range, is quite staggering.

b. Comparison is assisted if all diagrams start from a common base and their scales are the same. It is easier to compare the numbers in the different age groups in diagram 16.2 than in 16.1 because of this.

c. It should be made quite clear what the diagram is about, with an adequate title, labelling of axes of graphs etc. and scales clearly indicated. Often the actual data may be quoted as an addendum for the perusal of those whose attention is really captured.

d. The diagram should stand out clearly from its background, only the minimum framework of guide lines such as is found with graph paper being retained. Heavily printed graph paper is best avoided for display purposes. When special scales, such as logarithmic are used, this should be clearly indicated.

e. Colour is a great help in making a satisfactory diagram though it increases the cost if printing is being considered. Shading may be a partial solution of this.

f. The relative scales on vertical and horizontal axes can result in a graph being stretched upwards or sideways and a misleading impression may be created. If we wish to emphasise the absolute variations we may use a large vertical scale and show only the top of the graph. In such cases a break should be incorporated in the vertical axis (usually shown by a wavy line) to indicate that a large part of the graph is missing. If this is not done a person hastily viewing the diagram might interpret it as showing much greater relative variations than is really the case.

The remainder of this chapter is devoted to examples of some special forms of diagram used to illustrate statistics of various kinds.

## 16.3 BAR CHARTS

Examples are shown in diagrams 16.1, 16.2 and 16.3. They require little in the way of explanation, which is always a good point where drawings are concerned.

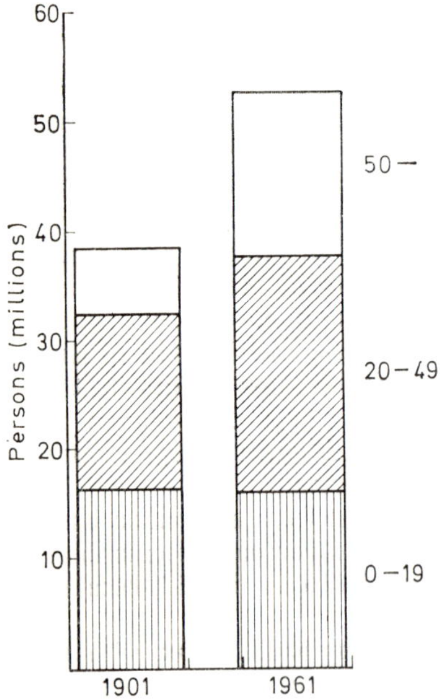


Diagram 16.1. *Bar charts of U.K. population*

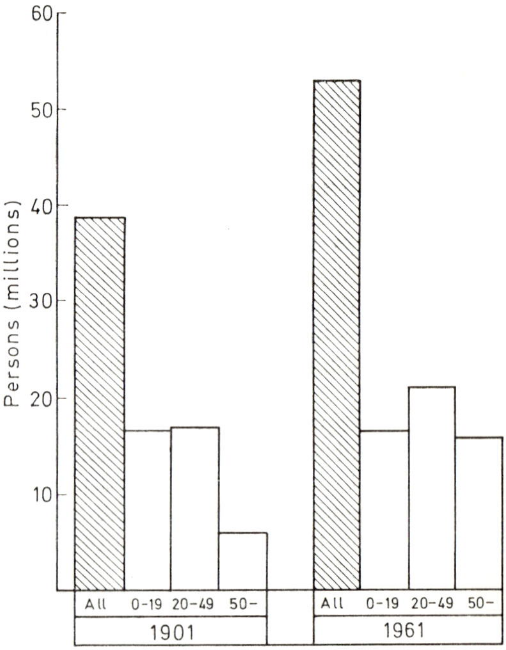


Diagram 16.2. *Alternative form of bar charts of U.K. population*

Diagrams 16.1 and 16.2 illustrate the same data. Notice how it is easier to make comparisons between the *components* of 16.2 than of 16.1, and also to compare each with its total. This is because they all have the same base. An alternative is to make each total bar the same height, regarding it as 100%, and to convert each of the sections from which it is built up into a percentage. This calls attention to changes in the pattern of distribution within the total.

The *population pyramid* (diagram 16.3) is an example of a popular adaptation of the bar chart to give a two-fold comparison. Not only can we compare age distribution but we can study also the balance of the sexes at different ages.

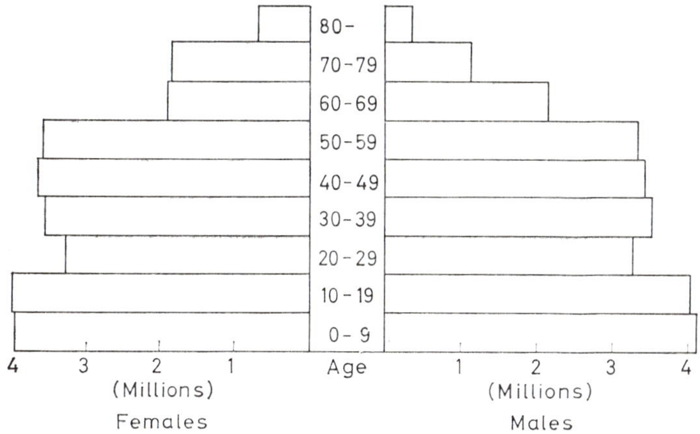


Diagram 16.3. *Population pyramid of U.K. population* 1961

A development of the bar chart, especially suited to the comparison of a total and its components over several periods, is to graph the total on a time base and also graph its components. The latter, generally shaded, are placed one above the other, so that they form a series of *bands* dividing the area between the base and the total line.

## 16.4 PIE CHARTS

These offer an alternative to bar charts when the composition of a total is to be illustrated.

The circle on the right of diagram 16.4 represents twice the total sales of that on the left. Since the area of a circle is $\pi \times (\text{radius})^2$ its radius is $\sqrt{2}$ times the radius of the left-hand picture.

The 360 degrees of the circle are divided in proportion to the relative size of the particular component. To avoid cumulative error in drawing it is best to calculate the cumulative number of degrees to each successive component, working from some radius that is regarded as zero.

The drawbacks of non-linear charts from the point of view of interpretation have been mentioned.

In the diagram it is easy to see that the proportion of labour in each total is the same, aided by the fact that both have a vertical radius as origin.

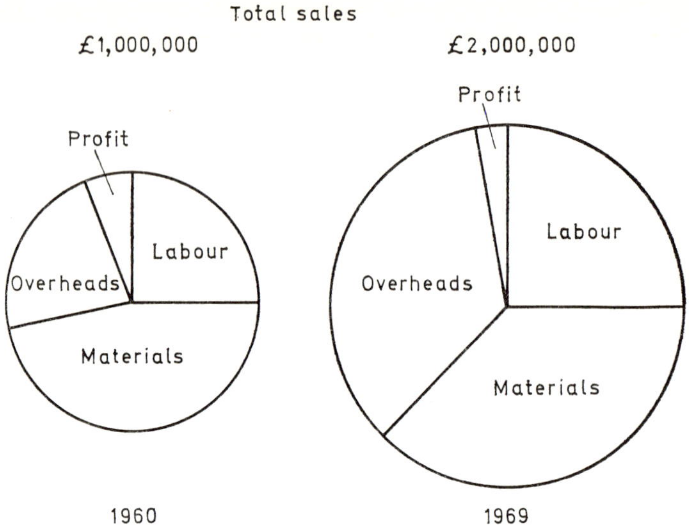


Diagram 16.4. *Pie chart* of sales revenue build-up

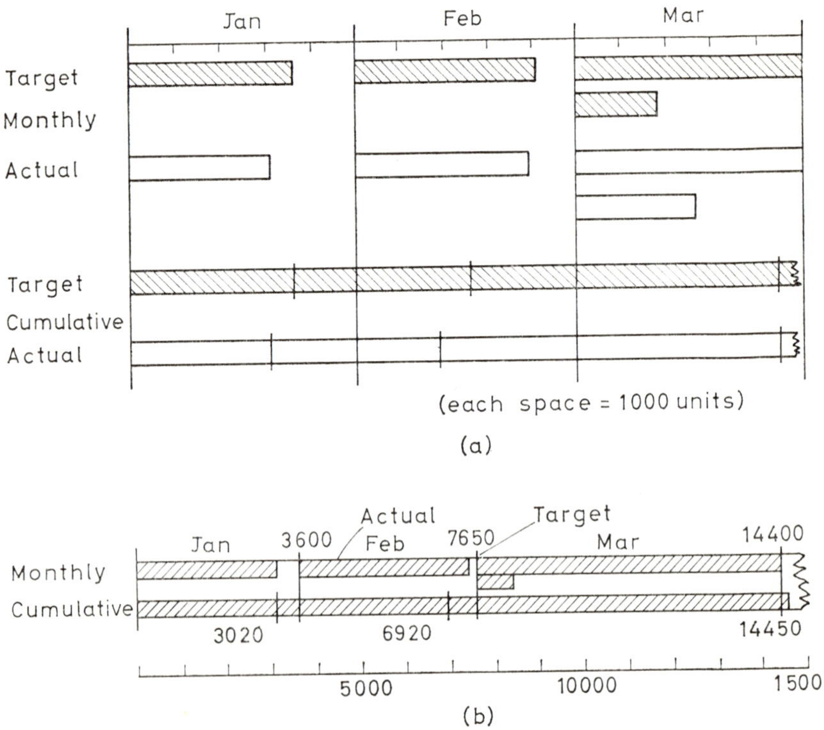


Diagram 16.5. *Gantt charts* for Example 15.4

Is it easy to judge that the absolute amount of profit in each case is the same, and that the percentage of overheads in the second case is half as much again as in the first, so that the absolute amount of overheads has increased to three times what it was?

## 16.5 GANTT CHARTS

These special bar diagrams (named after the American management pioneer) show how actual results compare with a target.

Diagram 16.5 shows two forms of this popular chart. In the upper part the chart is divided into equal spaces, each representing a month, in this case. Within these spaces target lines are drawn, first for the scheduled month's output and, lower down, for the cumulative output. If there is not enough room for a line it is continued immediately below the main line. The actual figures, as they arise, are similarly shown, and it can be easily seen how the subject matter of the chart is progressing.

A much handier version is shown in the lower part of the diagram. Here each horizontal space represents a monthly target, placed end-on with its predecessor, so that it shows the cumulative target as well. Within this bar are drawn thinner lines indicating actual and cumulative data.

An arresting version of the Gantt chart is the *thermometer* type of chart where a red line creeps up a drawing of a thermometer tube on which the target is indicated by a horizontal line.

In industry *machine load planners* are obtainable consisting of large boards with horizontal slots. Each slot indicates a separate machine. As jobs are scheduled for these machines duplicate job cards, their upper edges marked with a line proportional to the time to be taken, are placed in these slots, one behind the other, the left hand edges touching the end of the time line drawn on the previous card. The result is a line indicating how far ahead the machine has been loaded with work. As jobs go on to machines the top card goes to the worker concerned, with appropriate instructions, the lower card, in a different colour, clearly showing how the machine is occupied. The whole chart has a horizontal time scale along which a vertical wire can be moved to indicate the current time, showing work that is behind schedule.

An ingenious way of combining the linear properties of bar charts with the appeal of pictures is to indicate a number by small drawings, each of the same size, placed side by side. Each little man, say, represents perhaps a million workers, so that six men linked together would denote six million. This is not very accurate for fractions of the unit, and half men look rather odd, but careful selection of the unit size can reduce these defects.

## 16.6 PLANNING NETWORKS

The network form of planning diagram has become very popular. A project is regarded as consisting of *events*, which are stages in its completion, indicated by numbered circles, and *activities*, which lead from one event to the next and involve expenditure of time and resources. *Dummy activities*, shown by broken lines, are used to indicate the dependence of one event on another where time and resources are not involved.

Anything like a full treatment of this subject is beyond the scope of this book but table 16.1 shows some typical data. It relates to the replacing of a greenhouse with a new model ordered by post and has the same numbered activities and times as table 8.1 where some of the statistical aspects were considered.

Table 16.1. *Replacement of greenhouse*

| Activity | Description | Estimated Time in Days |
|---|---|---|
| 1—2 | Obtain brochures etc. | 14 |
| 2—3 | Decide on model | 4 |
| 3—4 | Demolish old greenhouse | 3 |
| 3—5 | Place order | 3 |
| 4—6 | Dig new foundations | 5 |
| 6—7 | Prepare for erection | 8 |
| 5—7 | Await delivery | 29 |
| 7—8 | Erect greenhouse | 7 |

The network is shown in diagram 16.6, in which the longest path (the *critical path*) has been shown by a double line. Such a diagram is not drawn to scale but is intended to show the way in which the whole project hangs together. In the lower part of the diagram is the bar-chart equivalent of the same data.

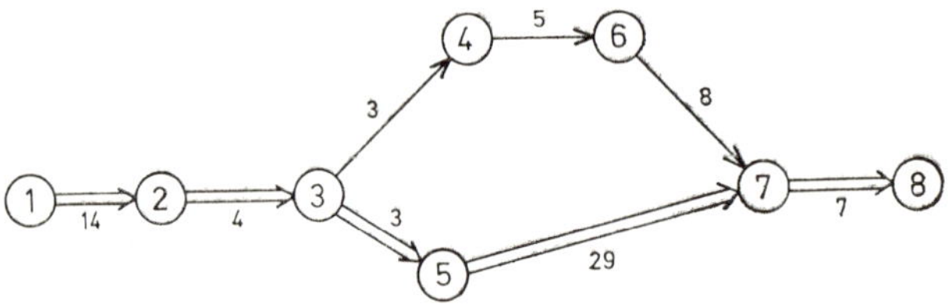


a Network diagram (not to scale)

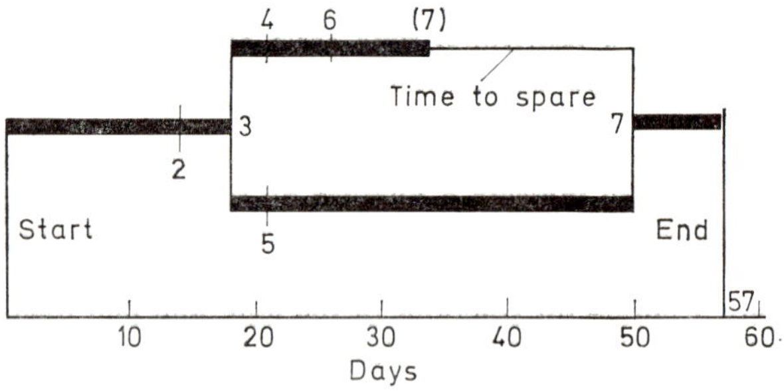


b Bar chart (to scale)

Diagram 16.6. *Planning diagrams*

The sequence in which activities are listed in a table such as 16.1 is not necessarily an indication of the order in which they should be started. A main purpose of the analysis of networks is to determine this and how much time there is to spare on paths other than the critical path.

## 16.7 Z-CHARTS

The idea of the Z-chart is to show three lots of planning data on the same chart:

1. Monthly (or weekly etc.) data
2. Cumulative data throughout the year
3. Moving annual totals throughout the year.

These are placed on the chart, first in the form of target lines, then as actual lines as the data becomes available, so that comparisons can be made and remedial action instituted at an early stage, if necessary.

The simplest form of the diagram, shown as an inset in diagram 16.7, has all the data on the same scale. The monthly targets appear along the

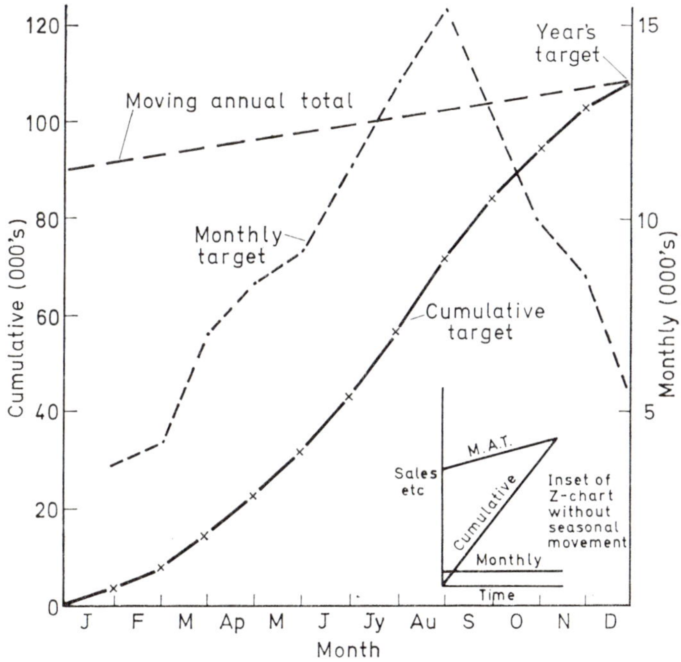


Diagram 16.7. *Z-chart*, Targets only, for Example 15.4, with seasonal variation. Inset is the simplest form of this chart

bottom, the cumulative target is a straight line connecting zero with the total year's target, and the M.A.T. line connects the previous years total to this latter point. The result is a form of diagram resembling the letter Z.

This suffers from two defects:

1. The scale for the monthly data is small. In the main diagram a separate scale has been used on the right.
2. The straight line cumulative target assumes that there is no seasonal movement in the data. If there is it should be shown, or the whole project is vitiated.

The main diagram shows the data from the example of the previous chapter (section 15.4, table 15.6).

## 16.8 WORKED EXAMPLE

*Question*

Suggest diagrams suited to display the following:

*a.* Comparison of the relative distributions in two countries of private cars, goods vehicles, and public service vehicles.

*b.* Deviations over a long period of the percentage unemployed in the working population about some level taken as standard.

*c.* The effect of an increase in the national income on the amounts going to the different factors of production.

*Answer*

[This answer would require neat sketches indicating the form of the proposed diagrams. These are not included here for reasons of space.]

*a.* A bar chart in the form of diagrams 16.1 or 16.2, with the data in the form of percentages of the totals for each country, which are then the same length, so that the emphasis is on the relative distributions.

*b.* An ordinary graph, showing the percentage unemployed vertically and the time horizontally. A horizontal line will indicate the standard level, and deviations from this can be emphasised by shading the area between the curve and this line.

*c.* A pie diagram for each level of national income would be a popular answer. A bar chart in the form suggested for part a. would be better, though perhaps less appealing. Perhaps the best solution would be a band chart.

## 16.9 PRACTICE QUESTIONS

1. Illustrate the data of table 6.1 by means of a suitable diagram.

2. Using the data of table 3.11, relating only to establishments with 2,000 or more workers, draw a diagram comparing the vehicle and textile industries.

3. Discuss the use of diagrams in the planning and control of business operations, mentioning specific types for particular purposes.

4. State some of the chief ways in which diagrams may mislead their viewers and suggest remedies.

5. Do you consider that the obvious popularity of pie charts justifies their use? Give reasons for your answer.

6. Write explanatory notes on three of the following: *a.* component bar chart, *b.* population pyramid, *c.* Gantt chart, *d.* band chart.

# 17

# Statistical Investigations

## 17.1.1 SURVEYS AND EXPERIMENTS

Statistical investigations are concerned with increasing our *knowledge of situations*, as they exist, or of indicating the *likely outcome of changes* that we may bring about.

In the *first* category, of investigation into existing facts or *surveys*, there are two main sectors of interest:

a. Research into economic and social aspects and into certain technical matters. Examples are afforded by market research, by opinion polls, by studies of management activity (such as how managers spend their time), by organisation and methods studies of the types of office equipment used by businesses and the work for which it is employed, and by operational research investigations into the arrival patterns of lorries at a loading point to provide the basic data for a queue study.

b. Provision of control information. Control implies the comparison of estimates and plans with the facts as they arise and is an essential part of management activity. The ultimate aim is to feed this information back, as quickly as possible, to enable appropriate management action to be taken. Production and cost control are important examples. Much of this is routine, but the statistician can give useful advice on the collection of data and how to interpret it, as in the case of statistical quality control (chapter 10).

In the *second* category, where we are concerned with estimating the likely effects of taking specific actions, we have *experiments*. Here the formulation and testing of hypotheses is important, and we have a basic philosophy of approach and rules for the practical design and conduct of experiments.

In practice it is often difficult to distinguish between surveys and experiments.

## 17.1.2 DESIGN OF EXPERIMENTS

The chief thing to bear in mind in connection with experiments, as distinct from a simple examination of an existing situation, is that the experimenter is trying to estimate the effect of some change introduced by the experiment. He is particularly concerned with eliminating the effect of changes in other factors, or he may admit them and try to estimate their contribution to the total effect.

If we wish to estimate the percentage of people who eat a certain item of confectionery our concern is to obtain a reliable sample (chapter 9). If we further wish to find if there is a significant difference between one part of the population and another, in this respect, we can take two samples and test any differences for significance. These are not really experiments.

An experiment would arise if we wished to test the effect of advertising, or of changes in wrapping and presentation, on the sales of the product. The simplest approach would be to establish the degree of simple correlation between the amount of advertising and the sales, allowing an appropriate time lag. A very long period of time would be needed to accumulate enough data to establish this degree of correlation. When it was established we should require information as to the part played by other factors. The main reason for the increase of sales may have been a general rise in spending power.

Experiments are carried out either to establish a theory, where none has previously existed, or to test, and perhaps modify, an existing theory or hypothesis. Several stages are involved in this:

1. Collection of data
2. Analysis of data
3. Formulation of a theory
4. Testing the theory.

In fact this is a continuous process. Stage 4 gives rise to the collection of fresh data, and so on.

Theory can be developed both by *induction*, where it arises from the facts, which may show a developing pattern, and by *deduction*, where the logical conclusions of certain hypotheses are studied. Both methods are used in practice, though researchers often show bias in favour of one or other of these approaches.

The questions with which the statistician may be involved are outlined in the last section of the present chapter. Before this general summary we shall consider some specific problems and approaches in the field of research and experiment.

Experiments vary in complexity. The simplest is concerned with the effect of varying a single factor. This involves the use of two groups, an *experimental group* in which the factor is varied, and a *control group* in which it is left constant.

We must take care that the two groups are basically the same. The problem is similar to that of sampling, and the groups may be made up randomly. It is most effective to have an equal number in each group.

If there is a significant difference between the two groups with respect to the result we are measuring we can reasonably attribute this to the change we have induced. The danger is that other causes may come into play and one or more extra groups may have to be used to check on this. Probably the best known example is in the case of testing the effects of some forms of medical treatment, where the psychological factor may be important, as well as the physiological aspect on which the experiment will probably be originally conceived. A third group is used which thinks it is receiving, say, the same drug as the test group, while in fact it is being treated with a placebo.

An extension of the random group approach to experiments is to use *random blocks*. Basically this is similar to stratified sampling. Each block consists of groups, basically similar, which are given different treatments.

A special terminology has evolved in connection with such experiments. It is couched in terms of agricultural experiments in which much of the development work was done. Each part of a block which receives a separate treatment is termed a *plot*. The changes that go to make up the treatment of each plot are termed *factors*, and the number of variations possible in each factor is termed the *level* of that factor. The assessment of the result on each plot is termed the *yield*.

Where we are concerned with assessing the effect of changes in more than one factor the question of factorial design arises.

The simplest design is *two-factor* design, of which an example is given in table 17.1. Each cell of the table would contain an equal number of items, chosen randomly. In this case there might be 20 individuals in each cell. The performance in each cell would be measured by the average score of the individuals in it.

Table 17.1. *Two factor design assessing the effect of teaching method and type of course*

| Teaching Method | Type of Course | | |
|---|---|---|---|
| | Full-time | Sandwich | Part-time |
| Lecture | | | |
| Project | | | |
| Case Study | | | |
| Combination | | | |

A development from this is the *Latin square* where a third factor is introduced indicated by letters A, B etc. This third factor occurs once only in each row and column. This implies that the tabulation must be square.

The Latin square seems to have been used originally in connection with agricultural experiments. Crops receive different treatments of, say, fertiliser and the different yields are compared. In order to check the effect, if any, of planting in different positions in the test grounds the field is divided into rows and columns, in each of which each of the treatments appears once and once only. Analysis of variance (chap. 20) can be used to estimate any variance arising from the position of rows or columns.

Table 17.2. *A Latin square*

| Plot | Block | | | |
|---|---|---|---|---|
| | 1 | 2 | 3 | 4 |
| 1 | A | B | D | C |
| 2 | C | D | B | A |
| 3 | B | A | C | D |
| 4 | D | C | A | B |

Table 17.2 shows a Latin square for this sort of situation. The letters would represent different treatments and the square can be read as a plan of the ground to be used. If we get away from the agricultural aspect the square no longer remains a ground plan, becoming instead a plan of the treatments that the subject of our experiment will receive. Each row will receive a different treatment of the first factor; each column will receive a different treatment of the second factor; each letter will receive a different treatment of the third factor.

If the experiment is not conducted on a scale sufficient to allow all the factors to be assessed in each block we may have to use a *balanced incomplete block* where each factor occurs to the same extent but all the factors are not fully covered in each block.

Analysis of the results of experiments may call for *analysis of variance* (chap. 20), simple, multiple or partial *correlation* (chap. 12), *factor analysis* (concerned with a development of correlation techniques where the situation is simplified in terms of new variables or factors) and *principal component analysis* which concentrates on the chief causative factors. The term *multi-variate analysis* is often used to cover the general situation of the connection between a result and several causes of which we wish to know the relative importance. Tests such as *contingency tests* and *chi-squared tests* (chap. 20) may be used for hypothesis testing.

### 17.2.1 SURVEY TECHNIQUES. MARKET RESEARCH

Market research has expanded very rapidly in recent years. This has probably been influenced by the increasingly accepted American view of business, especially the school of thought that regards marketing as the key to business policy, and the satisfaction of consumer needs as the chief justification for the existence of the business unit.

Broadly, market research is an investigation into all aspects of the market involving:

a. The purchasing public. This, often called *consumer research*, has become popularly accepted. Studies of who buys what, where and when it is bought, what people think of what they buy and how they use it, what they think they really want, and who really makes the decision with regard to purchase, can be useful as aids to sales policy, the design of products and the elimination of unsatisfactory lines.
b. The market for the product in the full economic sense of the term. This involves investigations into the share of the total market held by different products, the effectiveness of various channels of distribution, the effectiveness of advertising and sales promotion, and the policies and practice of rival manufacturers.
c. The background economic situation in which the market operates. Included here are studies of population trends, age and sex distributions of the population, changes in taste and fashion and the likely effect of economic policies.
d. Special studies of overseas markets. Basically this covers the same ground as 1–3 above, with the addition of studies of the impact of such things as import regulations, payment problems, and the effect of international politics.

Different problems will probably be met in the case of *consumer* goods compared with those encountered if the market for *capital goods* is being studied. One big difference is in the number of purchasers usually involved; another is that the capital goods demand is often derived from that for consumer goods.

### 17.2.2 SURVEY TECHNIQUES. ATTITUDE SURVEYS AND OPINION POLLS

These are being used increasingly in general sociological research and find a place in market research and in industrial sociology where, for instance, reactions of employees to their business organisation are of considerable interest.

An attitude can be treated statistically just as can a fact that is capable of independent confirmation, but there is no guarantee that it will not change with time.

A simple method of assessing attitudes is for the interviewer to use his judgment, after an interview of perhaps half an hour. This may give rise to observer bias, though considerable training should have aimed at reducing this.

Another approach is to have questions the responses to which can readily be related to the attitudes being tested. The simplest way is to ask the person to tick what he considers to be his attitude from a prepared list. A less direct method poses questions to which the replies can be allocated to a group favourable or unfavourable to the attitude being considered. This judgment is based on a prior assessment of the views of experts on these same questions (in the case of *Thurstone* scales), or of the views of a sample of the general public (*Lickert* scales).

The *Guttman* (G) scale, sometimes called *scalogram analysis*, is based on a cumulative scale. Agreement at any level implies agreement with any lower level. If you agree that something is more than 5 feet long you must also agree that it is more than 4 feet long.

The *scaling* process is the one generally used for evaluating attitudes. Probably the most popular scale is the *5-point scale* where the attitude must be summed up by one of 5 possible statements such as, *strongly disagree, disagree, undecided, agree, strongly agree*. These can be coded with numbers 1 to 5 for purposes of punched card or computer treatment. A *3-point scale* gives less choice, for example *no, undecided, yes*.

*Ranking scales* imply that statements are placed in order. The term *binary comparison* refers to the comparison of one item with the rest with a view to finding an extreme item. Repetition, with this item omitted, gives the next most extreme, and so on.

Ranking scales are usually examples of *ordinal scales*. They do not indicate amounts by which items differ and the difference may vary as between differing pairs of individuals the same 'distance' apart on the scale. *Interval scales* imply that, in addition to order, the difference between adjacent points is the same at different parts of the scale. Much psychological and sociological work uses ordinal and interval scales. Neither has the merit that the scale value is proportional to the absolute amount of the thing being measured, as is the case with *ratio scales* commonly met in the physical sciences. An intelligence quotient of 140 does not imply twice the intelligence of a quotient of 70, since there is no zero intelligence, but a weight of 140 lb. is twice as heavy as one of 70 lb.

There may be a difference between the attitude a person expresses and the attitude he seems to show, based on an objective study of what he does. Subterfuges, such as pretending that the objective of a survey is other than what it really is, may counter attempts by a person being interviewed to disguise his real attitude.

Attitude surveys usually require a closer involvement between the researcher and his subject than most other forms of statistical survey; interviews, often very time-consuming, may have to be used.

## 17.3 DATA COLLECTION. QUESTIONNAIRES. INTERVIEWS

From the point of view of cost and time involved the collection of data may be the most important part of statistical work, especially if the survey is an original one with no data available at the start. Very big savings can arise from careful planning of surveys, taking account of sampling theory and the sources and methods of presentation of the data. A *pilot survey* may be useful to indicate difficulties that may at first be unsuspected.

The main sources of statistics are listed below, together with comments on the practical problems involved.

a. *Published statistics*. Chapter 18 is devoted to these. The main problem is that such statistics are rarely precisely what we should like. They are either prepared to be of general use or are the outcome of some specific research having an aim different from our own. We must consider the universe to which they relate and the definitions of the units involved.

b. *Basic business data*. Any business is involved necessarily in a considerable amount of record keeping for the efficient running of its departments. Modern techniques, such as those using punched card equipment and computers, allow data to be easily processed in many ways. For example, financial accounting data may be used as the basis of sales analysis. The statistician can contribute his views on the design of documents. He may better appreciate the wider uses of data beyond the department immediately concerned. This may extend even beyond the particular firm, with an attempt to integrate relevant data concerning its customers and suppliers.

c. *Questionnaires*. These are lists of questions addressed to selected individuals who are thought to have the information we require. Where the individual is left to complete the document himself absolute clarity of instruction is needed, the questions should be unambiguous, and it is best that the answer should be a number or a choice between stated alternatives (for example—Yes/No/Do Not Know). Answers calling for a statement of opinion in the person's own words may cause great difficulty from the point of view of subsequent coding so that they can be treated statistically.

In some surveys a standard questionnaire may be completed with the guidance of a member of the research team who should have been carefully briefed.

Care is needed in the preparation of the questions that they are not structured in a way that tends to produce a particular answer and, also, that the full range of alternatives is available to the person completing the form.

It is possible to test the consistency of replies to different questions by insertion of check questions. An obvious case would be in an investigation into peoples' spending patterns where it is clear that a person cannot in a year spend more than his income together with his withdrawals from savings and any credit he has received. Sometimes these check items may be included in the questionnaire itself. Often they can be obtained from independent sources. With experience certain relationships tend to be established between parts of the data and any replies departing seriously from the established norms may be checked.

Data collected in this way, including documents such as time sheets, should always be treated with caution, especially if the person concerned has some interest in a particular conclusion being reached. One precaution is to have the information vetted by superiors, but this may be unduly charitable to the motives of these individuals. Even with perfect honesty there is a psychological tendency to *rationalise* a situation, implying that we tend to give 'facts' that will present our own actions in a favourable light.

d. *Experiments and direct observation.* These have the merit usually of being under the control of the research and statistical staff. Observation techniques include *participant observation* and *activity sampling*. The former arises chiefly in social surveys and involves a research worker joining the group that is to be observed. The latter involves observations at random intervals of, say, how an individual or machine is occupied. From a large number of such random observations the pattern of activities can be established by a frequency distribution, using the technique of section 2.4.1.

The keeping of *diaries*, by a manager or his secretary, to show how time is spent on different management activities, has been developed by researchers such as Rosemary Stewart.

e. *Interviews* which are particularly useful in social surveys involving the assessment of peoples' opinions and impressions. They can also be the source of other information, as when experts are interviewed by a Royal Commission which subsequently publishes a report.

The cost of this method of obtaining facts is relatively high, but much depends on the depth at which the interview is conducted and the length of time that is devoted to each interview. At one end of the scale we have *mass-interview surveys* of a superficial type; at the other end are *interviews in depth* which may last several days but, more often, occupy perhaps half an hour or an hour. They are usually exhausting as far as the interviewer is concerned and this sets a practical limit to the number that can be conducted in one day.

The degree to which an interview is *structured* can vary widely. We have *directed interviews* where the subject is asked specific questions or, at the least, is given guidance as to what should be discussed. The danger of interviewer bias is always present in these cases. On the other hand are the *non-directed* or *open-end interviews* where the subject is allowed to talk about anything he likes. The interviewer may take notes and it is useful to keep a count of the frequencies with which particular topics occur over a series of interviews, as an indication of what is currently regarded as most important.

Probably the least formal type of interview occurs when the investigator talks things over with people he meets, without their usually being aware of his purpose (the *informal interview*). He then attempts to write up his recollections and his conclusions from them. There is no real check on the validity of such findings but they sometimes have a freshness often lacking in more formal research, corresponding more to out normal way of gaining experience.

With the *focused interview* the person is provided with an article to read, or may be shown a film, and his response to this initial stimulus is assessed. The original intention was to test the effect of propaganda techniques. With data available from a large number of tests to provide norms it can be used to assess attitudes and strengths of feeling.

## 17.4 PRESENTATION OF DATA

Statistical investigations are not complete until their findings have been made available to the persons who are concerned with them and who may be involved in making decisions or taking action based on them. This is the final stage of the communication process of which investigations form a part and is assisted by:

a. The use of diagrams, of as simple a kind as possible, to overcome the average person's psychological reluctance to becoming involved with figures. Diagrams can also aid the quick appraisal of data by specialists.

b. Emphasising the important aspects of the data. Probably the best way of doing this is to establish a criterion of what is significant and to make this the basis of what may be a routine management report, eliminating the rest. Nobody expects plans to be achieved precisely, and it serves no purpose to worry about minor deviations. On the other hand, anything of major significance should be drawn to the attention of management immediately so that action can be taken. In management theory this is generally called the *exception principle*.

c. If tabulations of any size are used the eye should be guided with respect to rows and columns by appropriate rulings or by the use of breaks in the data every five or ten items. The use of italics to emphasise that certain figures are different from the rest, for example percentages in a table of actual figures, is helpful.

d. Clarity as to the subject matter of a tabulation or diagram is essential. It may be useful to include the sources of the information but this is, perhaps, more appropriate to research reports of the academic type than to the average sort of management report. Grading of detail, where the aims and conclusions are presented first, followed by increasingly basic data, is a solution to this problem. Different versions of a report, with differing degrees of detail, may save costs, and it may be well to have a highly condensed version circulated to a wide audience to keep them informed.

e. The choice of parameters and diagrams should reflect the statistician's ability to avoid misconceptions and ambiguities such as an average that hides important detail, or a graph that gives a mistaken impression of rate of growth.

f. Management statistics should be as up-to-date as possible with an indication of errors that this may involve, perhaps due to all the information not being available precisely.

## 17.5 THE PLACE OF THE STATISTICAL DEPARTMENT

A statistical department in a business organisation usually acts in a advisory capacity, providing, when required, a service for the other parts of the organisation. It helps line management and the departments exercising special functions when they need statistical information or when they become involved in using statistical methods.

The actual collection of data, especially of the routine kind, is normally the duty of the department which will use it. There will, for example, be production data, financial and costing data, sales data and data relating to personnel. The specialists who work in these departments are normally professionally trained to treat their own facts and figures in particular ways, which usually have at least the merit of being understandable throughout the particular profession concerned. Most of these specialists are coming to accept the view that something can be gained from consulting with a trained statistician with regard to the collection, treatment, and interpretation of their information. Sampling schemes can offer economies, standards may become more objective, and application of the ideas of variability and probability may be conducive to greater realism.

In some cases the statistical department may be responsible for the collection and issue of data. The best known example is probably the Central Statistical Office, set up originally to provide such a service for the Cabinet. Even when this is the case it is usually chiefly a matter of collating information from other sources, presenting it in a suitable fashion, and, perhaps, analysing it.

Big business, trade associations, trade unions, and political parties normally have information or intelligence departments. Much of their work is concerned with keeping abreast of the latest developments so that data can be provided for their managers when they are involved in negotiations, public utterance or policy decisions. In a sense these departments can be regarded as statistical ones, though the bulk of their staff will normally consist of economists, labour relations experts and other specialists who are not primarily statisticians.

There is a growing interest in *information theory* and attempts are being made to integrate all sources of information in a business unit. The computer has probably been the basic cause of such developments. The statistician should be intimately involved.

The specialist role of the statistician in business and government is intimately bound up with providing for management a consultancy service on statistical techniques.

Management is vitally concerned with decision making and likes to feel that statistics will help to illuminate the future rather than exhume the past. Despite this essentially decision-oriented attitude there is still a need for some exploratory statistical work to fill in the background against which decisions may eventually be made.

In advising on a proposed statistical investigation the following points may have to be considered:

a. The purpose of the investigation. What is the basic problem?
b. The precise nature of the information needed.
c. The information already available and the additional information to be obtained.

d. The degree of accuracy required.
e. The information collecting techniques to be used.
f. The sample sizes to be used.
g. The volume of work and its cost.
h. Will the information justify the cost?
i. The time the investigation will take. When are the results needed?
j. The snags that are likely to arise.
k. The staff and other resources, such as data processing facilities, which may be involved.
l. The techniques to be used for analysis of the data.
m. The form of presentation of the data.
n. The persons to whom the data shall be presented.
o. The further uses that might be made of the data.
p. Is there to be a single *ad hoc* investigation or are repeats involved, perhaps giving rise to regular statistics.

A statistical department often originates in a specific management need. This might be the establishment of statistical quality control. From this it may develop to be more widely accepted and consulted. Eventually it may initiate investigations of its own.

Modern management is dependent on communication to enable it to exercise effective control. Such communication is increasingly factual and there should be vast scope for the development of statistical services in the future.

In this development a statistical department can spread the appreciation of its function by providing training facilities for members of the staff of other departments. This training may range from background courses, to enable others to appreciate the philosophy and techniques of statistics, to training in the practical use of specific techniques, such as acceptance sampling. The training up of its own staff to provide statisticians for the future is also vital when trained statisticians are so rare.

The size and degree of specialisation of a statistical department depends largely on the size of the organisation in which it operates. The small firm may not be able to justify the employment of even one full-time specialist. It can, however, retain the services of a consultant or, perhaps, make use of the facilities of a co-operative research organisation of which it may be a member. The large firm may have some dozens of graduate statisticians, with supporting staff that may easily treble this number.

Perhaps the most interesting post-war development has been in the field of operational research, which provides mathematical techniques for studying practical problems arising in the operation of business and other enterprises. Much of this work requires a sound knowledge of statistics. In connection with business it is now common to regard statistical methods as a branch of the wider subject of operational research, and to include statisticians in operational research teams.

## 17.6 PRACTICE QUESTIONS

1. Show how statistics can be of help in the planning of experiments and in the analysis of the results.

2. Draw a $5 \times 5$ Latin square and indicate an experiment to which it might relate, stating the factors you propose to investigate.

3. A firm is considering marketing a new product. Consider the assistance it may obtain from market research in this project. Indicate at least the general nature of the product, which you may choose.

4. Suggest the uses which may be made of attitude surveys in industrial sociology.

5. Consider the problems of assessing prejudices people may have and suggest means of overcoming these problems in statistical investigations.

6. Produce a short questionnaire with the aim of investigating some topic in which you are interested.

7. Compare and contrast *a.* structured or directed interviews, *b.* open interviews and *c.* informal interviews, from the point of view of their suitability for different types of investigation.

8. A business employing 20,000 people in general engineering is considering setting up a statistical department. Prepare a draft report to the management on the uses it might make of such a department.

9. A personnel department is concerned with the apparent high rate of labour turnover in a firm. Devise an investigation which may help the management in an analysis of the problem.

# 18

# Statistical Sources

## 18.1 THE CENTRAL STATISTICAL OFFICE AND GENERAL STATISTICS

The C.S.O. is the main co-ordinating agency for British official statistics. It was created to provide information for the Cabinet during the second World War and has since developed as the main source of published statistics, though still located in the Cabinet Office and reporting to the Prime Minister.

In 1968 a survey control unit was set up in the C.S.O. to examine forms and questionnaires emanating from government departments and other public bodies. The aim is to eliminate unnecessary duplication and to ensure compatibility of definitions.

*Statistical News*, published quarterly by the C.S.O., provides regular information on developments in government statistics.

The chief publications of the C.S.O. are the *Annual Abstract of Statistics* and the *Monthly Digest of Statistics*. Students should make a point of consulting these valuable publications and becoming acquainted with the range of statistics included.

The *Annual Abstract*, published from 1946, replaced the Board of Trade *Statistical Abstract*, which provided statistics going back to the middle of the nineteenth century and which ceased publication during the War. The contents cover the following main subjects: area and climate, population and vital statistics, social conditions, education, labour, production, retail distribution, transport and communications, external trade, overseas finance, national income and expenditure, home finance, banking and insurance, prices, an index of sources, and appendices showing the Standard Regions and Conurbations and the Standard Industrial Classification (used for classifying establishments according to industry). The current volume approaches 400 pages in length.

The *Monthly Digest* has a somewhat similar scope, though restricted to data for which monthly or quarterly figures are available. These latter are generally shown for the current year to date, with similar seasonal data for one or more previous years and totals for years more distant.

A *Supplement of Definitions and Explanatory Notes* appears with each January issue of the Digest and gives detailed definitions of all the terms and units used.

An *Abstract of Regional Statistics*, covering the United Kingdom, has been available since 1965. There are also *Digests* relating to Scotland, Wales and Northern Ireland. The regional abstracts collect the data under the Standard Regions and Conurbations into which Great Britain is divided.

*Economic Trends* appears monthly giving a regular review of the economic situation and more detailed surveys at intervals. Useful graphs show the trend of data, generally over a period of 10 years. Various series of *New Contributions to Economic Statistics* have appeared relating to articles in earlier issues of Trends. A similar approach is planned for *Social Trends,* due to appear at the end of 1969.

In collaboration with the Bank of England (as well as with other government departments) the C.S.O. produces *Financial Statistics* each month. Details are given of the financial accounts of various sectors of the economy, central and local government borrowing, public corporations, banking and money supply, other financial institutions, companies, capital issues and stock exchange transactions, interest rates and security prices and overseas finance.

Annually appear *National Income and Expenditure* (Blue Book), *National Income and Balance of Payments* (White Paper), and *United Kingdom Balance of Payments* (Pink Book). A guide to national accounts is to be found in *National Accounts Statistics* where the description follows the arrangement of the tables in the National Income and Expenditure Blue Book.

A useful account of statistical employment in government service is contained in *Statisticians in the Government Service* obtainable from the Civil Service Commission without charge. In 1962 the Treasury produced *Government Statistical Services.*

A *List of Principal Statistical Series Available* is brought up to date at intervals.

Separate from the C.S.O. is the *Government Social Survey,* set up in 1941 under the old Ministry of Information. This became a separate Department in April 1967. Its function is to apply methods of survey research on behalf of all sections of government and to co-ordinate government survey research. There is a *List of Published Reports and Papers.*

## 18.2 GOVERNMENT DEPARTMENTS

Here we are concerned with more specialised statistics with more detail then in the summarised publications of the C.S.O.

Some of the chief of these are mentioned below. An inquiry of the appropriate ministry or department will generally secure assistance in tracking down statistical information from less well-known sources.

The *Employment and Productivity Gazette* (replacing the *Ministry of Labour Gazette* from June 1968) appears monthly. The statistical sections include regular monthly data relating to industrial analysis of employees, overtime and short-time, unemployment, placing work of the employment exchanges, stoppages, changes in basic rates of wages and hours of work, and retail prices. There are also statistical series covering employment, unemployment, vacancies, hours of work, wages and earnings. Other statistics are to be found in special articles.

The same department produces the quarterly *Statistics on Incomes, Prices, Employment and Production,* providing factual information for those engaged in industrial negotiation or arbitration. A *Technical Appendix* appears in the first issue of each year.

The *Family Expenditure Survey* produces annual statistics on how households in the U.K. spend their money, primarily to fix the weights for the Index of Retail Prices (see section 5.7).

The *Board of Trade Journal*, published weekly, is a key source of statistics relating to business and overseas trade. A summary of 'selected indicators' is included.

The Statistics Division of the Board of Trade prepares a monthly *Report on Overseas Trade* giving details of imports and exports by commodities and countries, and volume and value index numbers.

A *Business Monitor*, consisting of sheets of statistics, is available on annual subscription.

Periodically *Census of Production Reports* and *Census of Distribution Reports* have appeared.

From January 1969 a *Business Statistics Office*, staffed and run by the Board of Trade but directed by an interdepartmental committee chaired by the Director of the C.S.O., has been established as a development of the Board of Trade Census Office. It is intended to become the main agency for collecting and publishing industrial and commercial statistics. It will build and maintain a *Central Register of Businesses* to replace the registers held by different departments.

The Registrars General produce population and vital statistics for England and Wales, Scotland, and Northern Ireland. A population census normally takes place every ten years in years ending with a 1, though a sample census of one-tenth of the population was held in 1966. *Census Reports* are available with increasing detail as the vast mass of data is processed.

A yearly *Statistical Review* has appeared since 1921. It is in three parts consisting of medical tables, population tables and a commentary. There is also a quarterly *Return of Births, Deaths and Marriages.*

Among the other published statistics emanating from official sources may be mentioned many publications of the Treasury relating to public investment, government expenditure, the National Debt and Exchequer Returns and Estimates; the monthly Progress Report of the Department of Economic Affairs; the monthly Weather Reports of the Meteorological Office; the Criminal Statistics of the Home Office; the quarterly Local Housing Statistics of the Ministry of Housing and Local Government which also produces Local Government Financial Statistics; the Annual Report of the Ministry of Social Security; the Annual Report of the Ministry of Health; the Statistics of Education of the Department of Education and Science; the Ministry of Power Statistical Digest; the Annual Report of Her Majesty's Commissioners of Customs and Excise; the Report of the Chief Registrar of Friendly Societies; the annual publications of the Ministry of Transport relating to Passenger Transport, Highway Statistics and Road Accidents and the monthly Return of Vehicles Registered, as well as the quarterly British Railways Board Statistics; and the Agricultural Statistics of the Ministry of Agriculture, Fisheries and Food.

## 18.3 OTHER BRITISH SOURCES

a. Local Authority statistics. A good example is the *Annual Abstract of Greater London Statistics* produced from 1968 by the G.L.C.'s Research and Intelligence Unit.

b. Regular newspaper or magazine type publications such as the *Financial Times* and *The Economist*.

c. The quarterly journals of the main banks.
d. The annual reports of public companies and corporations.
e. Reviews such as the *National Institute Economic Review* (produced quarterly from 1959) and the *London and Cambridge Economic Service Bulletin.*
f. Special publications such as *Key Statistics* (Times Newspapers) and the *Statesman's Year Book.*
g. Research publications and books on special topics. As an example there are the *National Readership Surveys* of the Institute of Practitioners in Advertising.

## 18.4 OVERSEAS DATA

Many overseas governments produce statistics comparable with those available in Britain but probably the best sources of comparative statistics are the United Nations Organisation (U.N.O.), the European Economic Community (E.E.C.), the Organisation for Economic Co-operation and Development (O.E.C.D.), the General Agreement on Tariffs and Trade (G.A.T.T.), the European Free Trade Association (E.F.T.A.), the International Labour Office (I.L.O.) and the International Monetary Fund (I.M.F.).

U.N.O. produce the *Statistical Year Book* and the *Monthly Bulletin of Statistics.* Among its other publications may be listed the *World Trade Annual, The Year Book of International Trade Statistics, The Demographic Year Book, The Economic Bulletin for Europe, The Year Book of National Accounts Statistics* and the *Annual Report on Agriculture.*

E.E.C. has produced *National Accounts* and the *Monthly Graphs and Notes on the Economic Situation in the Community,* as well as the very handy pocket-sized *Basic Statistics of the Community,* which deals not only with the six countries of the Common Market but also with E.F.T.A., Canada, United States, Japan and U.S.S.R.

O.E.C.D. has produced *Main Economic Indicators* and *Industrial Production* (quarterly) with an informative *Historical Statistics* as a supplement. There are also available *Sources of Statistics for Market Research,* analysed into products and regions. Another publication of interest is *Demographic Trends 1965 to 1980.*

I.L.O. produce a *Year Book of Labour Statistics* and a quarterly *Bulletin of Labour Statistics.*

I.M.F. has a *Balance of Payments Year Book* and the monthly *International Financial Statistics.*

The above publications have the merit of being in English, an advantage shared by Commonwealth and United States publications, which are fairly accessible.

## 18.5 PRACTICE QUESTIONS

1. You have been asked by your management to produce a short report suggesting a nucleus of basic publications on statistics relating to the United Kingdom to be available in a small management library run by your firm. Give a draft of this report which should contain reasons supporting your choice.

2. Why do you consider the provision of up-to-date statistics has become so important in a modern economy?

3. Where might you expect to find statistics of the following?

   *a.* The numbers of people in different age groups in the population.

   *b.* Value of computers delivered to British users.

   *c.* Details of Trade Union membership.

   *d.* Probation orders issued.

   *e.* Imports of foodstuffs from the United States.

4. Outline some new sociological survey you consider would be useful and indicate any sources of published statistics you might wish to consult to help in carrying out this survey.

5. A business whose marketing has previously been local is considering expansion into other areas of the U.K. Give an account of any published statistics which might be consulted.

# 19

# Data Processing Equipment

## 19.1 INTRODUCTION

The practical application of statistical methods inevitably involves calculation, often of a monotonous, repetitive kind. Large masses of data may have to be collected, sorted and added before most of the calculations mentioned in this book can be performed. Manufacturers of data processing equipment serve the statistician well, providing equipment from pocket slide rules, priced in shillings, to computers in the thousands of pounds.

A full treatment of such equipment would be more appropriate to a book on office methods. Here we can only refer to the broad categories of equipment. The reader will find that most manufacturers have excellently illustrated and annotated sales literature as well as display rooms. Often the skilled use of special equipment requires intensive training and it is doubtful if the professional user of statistics would wish to train as a calculating machine operator, any more than he would wish to be a high speed typist. He should, however, be aware of the advantages and limitations of the various calculating techniques. He will probably be able to use a slide rule (see section 21.2) and a desk calculating machine. Increasingly he will find it helpful to have a knowledge of simple computer programming.

## 19.2 CALCULATING MACHINES

The main categories are:

a. *Adding/listing machines* which have the advantage that there is a record, in the form of a list on a roll of paper, of all the items handled. This can be useful for checking the accuracy of data input. Totals and sub-totals can be obtained. They can be regarded as reliable, if somewhat slow, means of adding items, and are restricted to addition.

b. *Non-listing machines* (of the *Comptometer* type) which keep no record of the data input. The latest total appears behind a horizontal row of windows. Many of these machines can be used for multiplication and division, after converting the data to decimal form, by depressing the pattern of keys making up the multiplicand the number of times equivalent to the units, tens etc. in the multiplier. Fully effective use of such machines requires a trained operator, though the average clerk can achieve useful work on them.

c. *Desk calculating machines*, primarily intended for multiplication and division. These may be of the type where a number is set by pulling levers round a barrel into appropriate positions, multiplication being achieved by turning a handle the required number of times as the carriage is traversed into the appropriate positions for units, tens etc. More recently *electronic* calculators have appeared performing similar or more complex calculations, the number setting and operating being usually by means of keys. These machines are not basically intended for straight addition.

d. *Special equipment* of the *accounting machine* type is useful for analysis of sales etc. and may sometimes be used for non-accounting data where columnar analysis is needed.

## 19.3 PUNCHED CARD SYSTEMS

In these systems a card normally having 80 columns is used to record data by means of holes punched in these columns, the vertical position indicating numbers from 0 to 9. In conjunction with two extra positions the whole alphabet can also be coded. The columns are grouped in *fields* wide enough to hold the largest number likely to be encountered, and referring to a particular aspect of the data.

Cards can be *sorted*, column by column, which is a very valuable feature to the statistician. Each card will relate to a separate item of the sample or universe, and the same cards can be arranged in different ways, grouping in terms of different characteristics as coded in the different fields.

The *tabulator* will list and total items as required, printing the results as directed. Its operation is controlled by instruction *panels* which are fitted into the machine. Greater scope is afforded by a *calculator* unit for operations such as multiplying the items in two columns together.

Other apparatus exists for *collating* data (merging cards), for punching cards automatically from pencil markings on original cards (*mark sensing*) and for *verifying* punching, though this latter does not guarantee against human error.

## 19.4 COMPUTERS

Computers are of two main types:

a. *Analog* (or analogue), which use some physical variable, often electrical voltage, to represent the quantities involved. They can be designed to solve specific problems. They are rarely, if ever, used for commercial work or for the types of calculations dealt with in this book. Control of industrial processes provides one of their main fields of application.

b. *Digital*, which break down every calculation into numerical digits. The *binary* system is generally used, where the base is 2, and every number is represented by a series of 0's or 1's. (For a fuller explanation see chapter 21). Such computers are ideally suited to statistical calculations, especially with complicated formulae and masses of data, when they can often perform in seconds what would involve hours or days of calculation by other methods.

Digital computers first began to come into general use in the late 1950's and have since become much more sophisticated, while at the same time being accepted and no longer regarded as esoteric equipment for a few expert mathematicians. They enormously increase man's power to handle data, and without them many of the applications of mathematics to business and the social sciences that characterise modern management and research would be impossible.

At the heart of a computer are:

a. The *arithmetic unit* and *co-ordinator* or *controller*, forming the *central processing unit*. The former carries out the arithmetic operations and has a *logical* section which deals with any decision-making operations affecting the sequence of execution of instructions. The latter controls the flow of information within the computer, sees that instructions are followed, and directs the flow from the input devices and to the output devices. It is the terrific speed with which this unit can perform complicated calculations or can cope with masses of data that makes the computer so valuable.

b. The *storage* or *memory* unit, without which the computer would be just a rapid calculator of very restricted application. This unit stores the program of instructions, as well as the data on which the computer is operating, and also intermediate results and final results prior to output. It can also store instructions that translate from a simple computer language to the instruction system the particular computer itself uses.

In addition to this main storage additional storage in the form of magnetic tapes, discs etc. can be coupled to the computer.

The computer is linked with the work-a-day world by *input* and *output* devices.

The former enable instructions and data to be read into the computer and may take many forms, such as punched cards, punched paper tape, magnetic tape, and contact through a teleprinter keyboard. Output may be through similar units or through a line printer. Devices which are linked directly to the computer are termed *on-line*, while those which are not so linked are *off-line*. Output may be through a paper tape punch which is on-line, but this would then have to be fed through a character printer before the output could be read by individuals in the normal way, and this would be an off-line operation. The speed of operations of devices linked with the computer will govern its effective speed of operation.

It is the *program* that determines what the computer will do and it will be followed blindly so it must be absolutely correct. The early computers had to be programmed in a machine language appropriate to their design but now we have standard high-level languages which are simple to learn and can be used with different computers. Two of especial interest to statisticians are ALGOL (ALGOrithmic Language) and FORTRAN (FORmula TRANslation language). The former has no particular restriction as to the original layout of the program. The latter, based originally on punched card coding, requires to be written on a coding sheet with spaces representing the columns of such a card, each line representing a fresh card, and various zones of the card being restricted to specific purposes.

Before such a program can be run, whether it is in the form of paper tape or some other input, the appropriate *compiler* for that language must be fed into the computer so that the latter can convert from the language used.

Computer design and applications advance very rapidly. The trend is to more powerful machines, to equipment that can be added to as a greater volume of work arises, to central installations which process the data for outlying branches, and the sharing of smaller firms in one computer. An important development is the input of data as it arises (*real-time* computing) coupled with switching from one program to another. In the field of *software* (that is programs etc. as distinct from *hardware*, the machines themselves) rapid developments are taking place.

Desk computers have appeared in which programs can be inserted in the form of magnetised cards. These seem handy for fairly straightforward statistical programs.

## 19.5 A SIMPLE ALGOL PROGRAM

```
ARITHMETIC AVERAGE;
"BEGIN"
"REAL" X,T,M;
"INTEGER" N;
N:=0; T:=0;
AA: "READ" X;
    "IF" X<O "THEN" "GO TO" BB;
    T:=T+X;
    N:=N+1;
    "GO TO" AA;
BB: M:=T/N;
"PRINT" M
"END";
```

In the above the first line is the title of the program, which calculates the arithmetic average of any number of positive numbers read in.

The letters after *real* indicate that these three symbols have been declared to represent real numbers (having decimal parts, if necessary) representing the number read in each time (X), the total up to this number (T) and the required average (M). The letter N, after *integer*, is declared to be a whole number, and represents the count of the items read in.

The following line sets N and T initially at zero, which is required since instructions subsequently accumulate anything which may be in this part of the memory.

*AA* is a guide to where we shall subsequently loop back into the program. The instruction is to *read* X. The value of X just read in is added to the accumulated total T (note that *:=*implies *becomes* not *equals*), and 1 is added to N to give the count of the number of items. The instruction is then to go back to *AA* and read the next number, and so on.

The input data will have a minus sign before an artificial last item so that when this is reached it is recognised by the computer, which goes directly to the final calculation at *BB*. This instructs the computer to divide the total of the items by the number of items and make M become this.

Finally this value of M is printed and the program is at an end.

There are slightly different forms of Algol, as of the other languages, so care must be taken that the form used suits the particular compiler.

## 19.6 MANUAL SORTING

At the other end of the scale from computers are the needle sorting systems for relatively small amounts of data. One system uses cards with holes pre-punched round the edges. Each position signifies an item of information, such as male or female, or a number. The existance of this particular feature is indicated by clipping away more of the card so that the hole is extended to its edge. The cards are lined up and a rod is inserted through the hole on which we are sorting; this supports all the cards except those which have this particular piece of information. An extension of this system uses cards which are pre-punched over an area, the space between different holes being clipped away. A special holder is used for the sorting.

A reversal of approach is to have a card for each feature that is of interest. Different positions indicate specific items, a hole being punched if the item has that feature. If several such feature cards are stacked one behind the other then the light will be seen through all the positions denoting items which possess all these features.

## 19.7 PRACTICE QUESTIONS

1. Compare the facilities which might be available for statistical calculations in small, medium, and large business organisations.
2. Discuss the uses to which a digital computer might lend itself in the field of sociological research.
3. Draw a diagram allocating the 80 columns of a punched card to different fields suited to some statistical investigation which you may choose and brief details of which should be included in your answer.
4. Prepare a scheme of instructions to eliminate errors in the calculation of totals from many items of data. Two comptometers are available with operators.
5. Explain the basic differences between punched card equipment and digital computers, showing why the latter are becoming so popular.
6. Distinguish between the following pairs in connection with computers:
   a. on-line and off-line units
   *b.* binary and denary (or decimal) numbers
   *c.* hardware and software.

# 20

# Further Techniques

## 20.1 ANALYSIS OF VARIANCE

The reader should be thoroughly conversant with the subject matter of chapter 6, relating to the standard deviation and variance, and also that of chapter 17, in which the design of experiments is discussed.

In section 6.5.6 we saw that variance has the property of being *additive*. The variances of different samples from the same population were added, after being suitably weighted by multiplying by the number in each sample. Division of the total by the appropriate number of degrees of freedom gave an estimate of the population variance.

This idea can be extended to show the analysis of the overall variance of data (when all the items are considered individually) into variance within groups into which the data is divided, and variance between the averages of these groups.

If an experiment is being conducted the different groups might receive different treatments, and we should be interested in how much of the total variability could be attributed to these treatments, which would reveal itself in the variance between the group means.

When doing such analysis it is usually best to work in terms of squared deviations from the appropriate means. These can then be converted into variances by dividing by the appropriate degrees of freedom.

A useful approach is to square the actual items involved and adjust the total of the squared values by deducting an amount equal to $\frac{(\text{sum of the items})^2}{n}$, which is the same as $(n\bar{x})^2$.

If we have the numbers 8, 15, 17 and 20, the squares are 64, 225, 289 and 400, of which the sum is 978. The items themselves total 60. The sum of the squares of deviations from the mean is thus $978-\frac{60^2}{4}=78$. To check this, using the actual deviations, we have the average of the numbers$=\frac{60}{4}=15$, so that the deviations are $-7$, 0, 2 and 5. The squares of these deviations are 49, 0, 4 and 25, of which the total is 78. This method is most useful when the mean does not work out exactly.

With some data the calculation can be simplified by dividing all the items by a common factor. A common amount can also be subtracted from all items, preferably about the size of the mean, so as to reduce the size of the figures involved. These adjustments, sometimes called *coding* the data, will alter the variances but will leave them in the same proportions, which is often our only concern.

We may use the bonus data of table 2.1 to simulate an experiment. Let us suppose that the 150 items have been divided into 6 groups of 25 items each and that these are represented by the columns.

Table 20.1. *Analysis of variance of bonus data (based on the 6 columns of table 2.1)*

| | Squared Deviation | Degrees of Freedom | Variance Estimate |
|---|---|---|---|
| Between items in each column from the column mean | | | |
| Column 1 | 21·58 | 24 | 0·90 |
| 2 | 17·71 | 24 | 0·74 |
| 3 | 9·17 | 24 | 0·38 |
| 4 | 5·79 | 24 | 0·24 |
| 5 | 12·72 | 24 | 0·53 |
| 6 | 19·22 | 24 | 0·80 |
| | 86·19 | 144 | |
| Between means of columns (×25) | 8·26 | 5 | 1·65 |
| All items from general mean | 94·45 | 149 | |

Table 20.1 gives the analysis of the variance of this data. The calculation of the squared deviations is not shown but may easily be checked. The only point to cause difficulty is the matter of the scale. It will be seen that the squared deviation between the means of the columns has been multiplied by 25, the number of items in each column, to make this value comparable with those dealing with individual items.

In this case there are 150—1=149 degrees of freedom overall. Of these 6—1=5 exist between the group means, and 25—1=24 exist between the items in each group, giving 24×6=144 degrees of freedom for the 6 groups. Division of the squared deviations by their appropriate degrees of freedom gives the estimate of the variance of the item being considered. It should be noted that these variances cannot be added directly. This must be done through the squared deviations.

In terms of experiments this is the equivalent of a *single factor* experiment, assuming that each group receives a different treatment.

Extension of this technique enables an analysis to be made of the results of experiments, the design of which has been dealt with in the chapter on statistical investigations (section 17.1.2).

If *two factors* are involved the total variance between the individual items can be split into:

a. that between the column means
b. that between the row means
c. the residual which is the difference between a+b and the total.

Table 20.2 gives the analysis for a *Latin square*. In addition to the columns and rows we have treatments, designated by A, B, C or D, appearing once in each row and column. The pattern of these treatments is the same as that in the specimen previously given in table 17.2.

The data have been squared, without prior coding. The lower part of the table shows the data relating to the treatments, which has been extracted from the upper part of the table.

Table 20.2. *Analysis of variance of a Latin square*

a. *Columns and Rows*

| Plot | Block | | | | | | | | Total | Av. | (Av.)² |
|---|---|---|---|---|---|---|---|---|---|---|---|
| | a | | b | | c | | d | | | | |
| 1 | A | 16 | B | 13 | D | 13 | C | 9 | 51 | 12·75 | 162·5 |
| 2 | C | 24 | D | 17 | B | 13 | A | 5 | 59 | 14·75 | 217·5 |
| 3 | B | 22 | A | 16 | C | 18 | D | 12 | 68 | 17·00 | 289·0 |
| 4 | D | 28 | C | 23 | A | 15 | B | 12 | 78 | 19·50 | 380·3 |
| Total | | 90 | | 69 | | 59 | | 38 | 256 | | 1049·3 |
| Av. | | 22·5 | | 17·25 | | 14·75 | | 9·5 | | | |
| (Av.)² | | 506·3 | | 297·6 | | 217·5 | | 90·3 | 1111·7 | | |

b. *Treatments*

| | A | B | C | D | |
|---|---|---|---|---|---|
| | 16 | 13 | 9 | 13 | |
| | 5 | 13 | 24 | 17 | |
| | 16 | 22 | 18 | 12 | |
| | 15 | 12 | 23 | 28 | |
| Total | 52 | 60 | 74 | 70 | |
| Av. | 13·0 | 15·0 | 18·5 | 17·5 | Total |
| (Av.)² | 169·0 | 225·0 | 342·3 | 306·3 | 1042·6 |

For columns, rows and treatments the sums of the squares of averages have been scaled to the equivalent of the individual items by multiplying by the number of items included, 4 in each case. This gives values of $1111{\cdot}7\times4=4446{\cdot}8$ for the columns, $1049{\cdot}3\times4=4197{\cdot}2$ for the rows and $1042{\cdot}6\times4=4170{\cdot}4$ for the treatments. The sum of the squares of all the items is 4624.

From each of these totals is subtracted the adjustment of $\frac{(\Sigma x)^2}{n}=\frac{256^2}{16}=4096$.

The results are the values of 350·8, 101·2, 74·4 and 528·0 shown in table 20·3. The residual is the balance.

Table 20.3. *Analysis of variance of a Latin square*

| Source | Sum of Squares | Degrees of Freedom | Variance |
|---|---|---|---|
| Columns | 350·8 | 3 | 116·8 |
| Rows | 101·2 | 3 | 33·7 |
| Treatments | 74·4 | 3 | 24·7 |
| Residual | 1·6 | 6 | 0·4 |
| Total | 528·0 | 15 | (35·2) |

Division of these figures by the appropriate degrees of freedom provides the variances required.

It will be seen that the largest contribution to the variance arises from the differences between the column means. The residual variance is, relatively, very small.

Instead of using averages for columns, rows and treatments we may work with the squares of the totals. In this case the respective sums will have to be divided by 4 to scale them down to the overall sum of the squares.

The F-test (section 20.5) may be used to test the significance of variations such as those just dealt with.

## 20.2 THE CHI-SQUARED DISTRIBUTION

We have seen in chapter 8 that the normal distribution can be used to test hypotheses by enabling us to estimate how frequently a given result would be likely to arise on a purely random basis from an assumed population. If it is quite likely then we attach no significance to the result. The parameters used in this test are the mean and the standard deviation.

Other parameters can be obtained which are of particular importance where the Normal curve does not apply. Of these the *chi-squared* test is perhaps the most important. It is used to test the significance of the differences between the frequencies of an actual observed distribution and the frequencies to be expected if the distribution obeyed strictly some stated law.

The Greek letter $\chi$ in its squared form is used (chi, pronounced ky), defined so that $\chi^2 = \sum \frac{(f_A - f_E)^2}{f_E}$, where $f_A$ represents the actual frequency of each item and $f_E$ its expected frequency. Sometimes it is easier to use the equivalent form $\chi^2 = \sum \left(\frac{f_A^2}{f_E}\right) - N$, which just involves squaring each actual frequency and dividing by its expected equivalent, summing the results and deducting from the total the total of the frequencies.

The test can be used as an indication of the goodness of fit of some theoretical distribution to actual data. The theoretical distribution may be in the form of a curve the equation of which is known, or, more simply, could be the expected distribution of frequencies in the cells of a contingency table (section 20.3).

Table 20.4. *Comparison of actual (observed) and expected frequencies*

| | Frequencies | | | | |
|---|---|---|---|---|---|
| X | Actual ($f_A$) | Expected ($f_E$) | $f_A - f_E$ | $(f_A - f_E)^2$ | $(f_A - f_E)^2/f_E$ |
| 1 | 18 | 22 | −4 | 16 | 0·7273 |
| 2 | 36 | 34 | −2 | 4 | 0·1176 |
| 3 | 50 | 48 | 2 | 4 | 0·0833 |
| 4 | 66 | 64 | 2 | 4 | 0·0625 |
| 5 | 81 | 82 | −1 | 1 | 0·0122 |
| 6 | 105 | 102 | 3 | 9 | 0·0882 |
| 7 | 114 | 124 | −10 | 100 | 0·8065 |
| 8 | 144 | 148 | −4 | 16 | 0·1081 |
| 9 | 179 | 174 | 5 | 25 | 0·1437 |
| 10 | 207 | 202 | 5 | 25 | 0·1238 |
| | 1000 | 1000 | | | 2·2732 |

Table 20.4 shows a typical calculation of the first type. The expected frequencies have been calculated from the equation $Y=12+9X+X^2$, which we may assume is the law being tested. The first column of values of X is not relevant to the calculation of $\chi^2$.

$\chi^2$ is seen to be 2·27 and we have to use tables of $\chi^2$ to find the probability of this value arising from chance, for the degrees of freedom associated with the data being tested. These are usually indicated in tables by the Greek $\nu$ (nu) and, in the present case, are $10-1=9$. If we work at the 5% level of significance the value of $\chi^2$ would have to be greater than 16·9 (from the tables). Thus we conclude that the difference between the actual and the expected values is not significant. In other words the actual distribution is a good approximation to the theoretical distribution.

It is possible to use $\chi^2$ as the basis of a measure of the degree of association by calculating the *coefficient of contingency* which, often denoted by C, is given from $\sqrt{\dfrac{\chi^2}{\chi^2+N}}$. This becomes larger the greater the association, becoming a maximum of $\sqrt{\dfrac{k-1}{k}}$ in a $k \times k$ table.

*Yates correction* is applied to the calculation of $\chi^2$ when there is only one degree of freedom, as occurs in a $2\times2$ contingency table (see next section) and when N is small. It consists of subtracting 0·5 (independent of the data) from the differences, before these are squared, with the object of approximating to the continuous variation of the normally distributed variable which is assumed by the $\chi^2$ test.

Care should be taken, before calculating $\chi^2$, that both the actual and expected total frequencies are the same. If not, they can be made so by expressing each column in percentages.

## 20.3 CONTINGENCY TABLES

The term *contingency table* is applied to a table of *frequencies* where there is more than one column and more than one row. It is usual to refer to an $h \times k$ table to denote one having h rows and k columns. The rows are always denoted first.

We are often concerned with testing whether the distribution of observed frequencies in the various cells of the table accords with the expected distribution arrived at by applying some hypothesis.

The simplest table is a $2\times2$ table such as is shown in table 20.5.

The upper half shows the actual distribution of passes and failures of the students of two colleges. We wish to test whether their is a significant difference between the colleges with regard to their rates of success.

If there is no difference between the colleges in this respect we should expect those who pass to be present in the same proportion in each of the columns. This proportion can be taken as that of the combined population, shown in the *total* column. Out of the total of 90 students 54 pass, which is 60%. Applying this percentage to the totals of each college individually we have 60% of 40 and 60% of 50. These are the expected passes which are entered in the lower table. The other expected figures can be filled in to make up the required totals.

In fact, with a $2\times2$ table we require only one of the four cells to be filled in to enable us to complete the rest, assuming the totals are given. A $2\times2$ table has, thus, only one degree of freedom.

Table 20.5. *Analysis of the performance of students of two colleges in selected examinations*

a. *Actual*

| | College A | College B | Total |
|---|---|---|---|
| Pass | 28 | 26 | 54 |
| Fail | 12 | 24 | 36 |
| | 40 | 50 | 90 |

b. *Expected, if there is no difference between the colleges*

| | A | B | Total |
|---|---|---|---|
| Pass | 24 | 30 | 54 |
| Fail | 16 | 20 | 36 |
| | 40 | 50 | 90 |

A simple trial will show that, for an $h\times k$ table the number of degrees of freedom is $(h-1)\ (k-1)$.

The chi-squared test can be applied directly to the contingency table, comparing the actual and expected frequencies.

For table 20.5 we have $\chi^2=\frac{(28-24-0{\cdot}5)^2}{24}+\frac{(26-30-0{\cdot}5)^2}{30}+\frac{(12-16-0{\cdot}5)^2}{16}+\frac{(24-20-0{\cdot}5)^2}{20}=3{\cdot}06$. Yates' correction has been applied, since there is only one degree of freedom.

Tables of $\chi^2$ show that, for one degree of freedom, this value would occur about 9% of the time, so the data cannot be regarded as indicating a significant difference between colleges A and B.

An alternative approach would be along the lines of chapter 9, treating the two columns as separate samples, and testing whether they could reasonably have come from the same population.

The calculation is as follows:

Proportion of passes for A and B together $=\frac{54}{90}=0{\cdot}6$

Proportion of passes for A $=\frac{28}{40}=0{\cdot}70$

Proportion of passes for B $=\frac{26}{50}=0{\cdot}52$

Difference between samples $=0{\cdot}70-0{\cdot}52=0{\cdot}18$

Standard error of this difference $=\sqrt{(0{\cdot}6\times0{\cdot}4)\left(\frac{1}{40}+\frac{1}{50}\right)}$
$=0{\cdot}104$

Difference $\div$ s.e. of difference $=\frac{0{\cdot}18}{0{\cdot}104}=1{\cdot}73$,

which, being less than 2, is not significant at the 5% level.

## 20.4 STUDENT'S t-TEST

This test (named after the pseudonym under which its discoverer wrote) is useful when we are dealing with small samples.

*Student's t* is defined as $t=\frac{(\bar{x}-\mu)\ \sqrt{n-1}^{*}}{s}$, where $\bar{x}$ is the mean of the sample, $\mu$ is the mean of the population, s is the standard deviation of the sample, and n is the number of items in the sample.

The population must be approximately normally distributed. When n is large, t is also distributed approximately normally, but for small samples this no longer applies, and the values of *t* associated with any given probability become progressively greater than if *t* were normally distributed. Application of a test based on the normal curve will tend to show significance at smaller differences than would be required by the t-test.

Tables of the t-distribution show the degrees of freedom ($\nu$ or nu) associated with the sample and selected probability levels. The former will be $n-1$ and the latter is often taken as 5% as the lowest level at which we can regard a difference as significant. As with the normal curve, levels of, say, 1% or less might be regarded as highly significant. If the value of *t* we have calculated is greater than the value shown in the table it indicates that the difference between the means being considered is significant.

It should be noted that some versions of the t-distribution table apply to the upper tail, while others refer to both tails. The probability level in the former must be doubled to give the same result as the latter.

When the population mean ($\mu$) is known, the test is very straightforward. The sample mean and the sample standard deviation (uncorrected) are used directly in the formula.

In cases where we wish to test whether two samples are from the same population we must first estimate the population variance by combining the two variances. This gives $\hat{\sigma}^2=\frac{n_1 s_1^2+n_2 s_2^2}{n_1+n_2-2}$

Next we estimate the standard error of the difference between the means, resulting from $\hat{\sigma}\sqrt{\frac{1}{n_1}+\frac{1}{n_2}}$.

Student's *t* is then $(\bar{x}_1-\bar{x}_2)\div$standard error of difference. The last step is to see if this is greater than the value in the tables for $n_1+n_2-2$ degrees of freedom and the level of probability we have chosen. If it is, then there is a significant difference between the two samples.

This test may also be used to test whether a coefficient of correlation obtained from a small sample differs significantly from zero. We use the fact that $t=\frac{r\sqrt{n-2}}{\sqrt{1-r^2}}$.

In cases where we wish to test if the coefficient differs significantly from some value other than zero we should use the Z transformation (see section 20.6).

---

* or $\left(\frac{n-\mu}{\hat{s}}\right)\sqrt{n}$, if Bessel's correction has been applied to s.

## 20.5 THE F-TEST

This is a test of whether there is a significant difference between various estimates of population variance.

The *variance ratio* (F) is the ratio between two estimates of this variance, the larger always being used as the numerator, so that F is always greater than unity.

In the case of two samples with variances $s_1^2$ and $s_2^2$ and degrees of freedom of $(n_1-1)$ and $(n_2-1)$, respectively, then $F=s_1^2/s_2^2$ (if $s_1>s_2$) or $s_2^2/s_1^2$ (if $s_2>s_1$).

If we wish to apply the test to variance analysis we can consider the ratio $\frac{\text{variance between groups}}{\text{variance within groups}}$ to estimate whether differences between different groups are due to variations arising in the groups or to real variations between the groups.

Tables of the F-distribution show degrees of freedom of the larger of the two variances as column headings, and degrees of freedom of the lesser variance as row designations. At the intersection of the appropriate column and row is the value of F which is significant at the level to which the table relates. If the calculated value of F is greater than this it indicates a significant difference.

## 20.6 THE Z-TRANSFORMATION

The *Z transformation* is useful for significance testing when the sampling distribution involved is not symmetrical. We convert the statistic into a Z value, which is approximately normally distributed, so that we can use probability tables of the normal curve.

This transformation is suited to testing the significance of the difference between a sample value of a correlation coefficient and some assumed value for the population, or between two different sample values. If we are just concerned with whether the value differs significantly from zero we can use the test given in section 12.6 or the t-test (section 20.4) if the sample is small.

Tables are available to convert any value of *r* into the equivalent value of Z. If these are not available we can use the formula $Z=0{\cdot}5\log_e\left(\frac{1+r}{1-r}\right)=1{\cdot}1513\log_{10}\left(\frac{1+r}{1+r}\right)$ and the fact that $\sigma_z=\frac{1}{\sqrt{n-3}}$.

For comparing with an assumed population value we transform r for the sample and $\rho$ for the population ($\rho$ replacing r in the formula), as above. We then find the difference between these transformed values, divide it by $\sigma_z$ and see if the result is significant, using the normal probability tables.

If two samples are to be compared the procedure is similar. We find Z for each value of r. The difference between these Z values is divided by an estimate of the standard error obtained from $\sqrt{\frac{1}{n_1-3}+\frac{1}{n_2-3}}$.

## 20.7 SIMPLE NON-PARAMETRIC METHODS

Most of this chapter, and indeed the bulk of this book, has been directed to interpretation and testing the significance of statistics.

There are broadly two approaches:

a. Assume that the population from which the data are drawn is distributed in some established way, which, generally, means normally distributed (i.e. following the Gaussian or Normal curve or error).

b. Make no assumptions about the nature of the population but rely on the observations themselves, which must be random and independent.

The second approach gives rise to *non-parametric* or *distribution-free* methods. These are often quicker to apply and appeal more readily to the common sense than some of the older established techniques. They are often regarded as *short-cut* methods.

Generally, where *measures* are involved (means, standard deviations etc.), the first, or parametric approach is desirable. When the data is in the form of *ranks* or of *counts* (frequencies, medians, rank differences etc.) the non-parametric techniques may be most appropriate. We can switch from parametric to non-parametric methods by simplifying the data, e.g. representing it by its rank rather than its measure. This usually involves discarding some of the information, so that the test may be less precise.

Some of the simplest of the non-parametric methods involve *sign tests*.

Suppose there are n pairs of items the scores of which are compared by placing a + or − sign in the adjacent column, depending on whether the first of each pair is greater or smaller. If these differences arise from chance there will be $\frac{n}{2}$ signs which are plus, and $\frac{n}{2}$ which are minus, in the long-run of sampling. We may say that the probability (p) of + is $\frac{1}{2}$, and the probability (q) of 'not plus', i.e. of − is also $\frac{1}{2}$.

The significance of any actual number of + items may be tested as an attribute, using the fact that the standard error is $\sqrt{npq}$, and that twice this is at the 5% level of significance.

Table 20.6. *Scores of 9 people in tests x and y*

| Person | Test x | Test y | Score Comparison |
|---|---|---|---|
| A | 36 | 34 | + |
| B | 39 | 41 | − |
| C | 35 | 33 | + |
| D | 38 | 37 | + |
| E | 41 | 42 | − |
| F | 42 | 39 | + |
| G | 37 | 35 | + |
| H | 37 | 34 | + |
| I | 40 | 36 | + |

In the example (table 20.6) there are 9 pairs (n=9), of which 7 show the first value as being greater than the second (7 plus signs).

If we assume that the + and − are equally likely to occur then the average number of + signs should be $9\times\frac{1}{2}=4\frac{1}{2}$, and the standard error should be $\sqrt{9\times\frac{1}{2}\times\frac{1}{2}}=1\frac{1}{2}$. At the 5% confidence level we are concerned with deviations of more than about twice this value, i.e. $2\times1\frac{1}{2}=3$. We should expect 95% of samples to be in the range 4·5−3 to 4·5+3, i.e. 1·5 to 7·5. Our value of

7 lies within this, so we are not justified in saying that the scores differ significantly.

The amount to be added to the mean expected number will always be $2\sqrt{n \times \frac{1}{2} \times \frac{1}{2}} = \sqrt{n}$, which provides a very simple test.

There are several versions of this test.

Obviously, we may use either + or — as the basis. In this case, if we work on the sign having the smallest frequency, we reach the same conclusion. The smallest number expected is 4·5—3=1·5. Since 2 is greater than 1·5, which implies it is nearer to the mean of 4·5, it is not significant.

Another version is to work on the *difference between the number of + and — signs*, regarding the two sets of data as varying significantly if this difference in number exceeds $2\sqrt{n}$. This method is based on the fact that the difference between the possible numbers of + and — signs rises at twice the rate of either one of them. This is shown in table 20.7 based on 6 pairs of items. The + items fall from 6 to 0, while the differences fall from +6 to —6. Since the average of the differences is zero we may apply the $2\sqrt{n}$ test directly, without calculating the average number of + items to be expected.

Table 20.7. *Possible combinations of + and — for 6 comparisons*

| + | — | Difference |
|---|---|---|
| 6 | 0 | 6 |
| 5 | 1 | 4 |
| 4 | 2 | 2 |
| 3 | 3 | 0 |
| 2 | 4 | −2 |
| 1 | 5 | −4 |
| 0 | 6 | −6 |

With the 9 items of the previous example n=9 and $2\sqrt{n}=6$. There are 7 plus and 2 minus signs, the difference being 5. This is less than 6 so the scores do not differ significantly.

An extension of this approach may be used to test the significance of the differences between *sample averages and a population average*. We take each item of the sample and compare it (+ or —) with this population average. If the two averages are the same + and — should occur with equal frequency The value of n is the number of items in the sample, and any of the above versions may be used.

If we take the method last described to test whether the items 52·3, 50·4, 51·6, 52·0, 50·1, 53·1, 51·8, 51·1, 50·8, 51·2, 50·8, 49·9, 51·9, 52·6, 51·8 and 52·6 vary significantly from a population mean of 50·5 we find that, of the 16 items (n), 13 are greater than this value and 3 are less. Since $2\sqrt{16}=8$, and 13—3=10, which exceeds 8, we may say that there is a significant difference.

This test may be used to indicate a significant *change in time series data*, comparing the later items with the average of the earlier ones, again on a plus and minus basis.

## 20.8.1 WORKED EXAMPLE

*Question*

*a.* A group of 12 children are found to have the following intelligence quotients: 112, 109, 125, 113, 116, 112, 123, 108, 113, 122, and 128. Is it reasonable to suppose that these children have come from a larger population whose average IQ is 115?

*b.* A second group of 10 children is tested, resulting in the following IQ's: 117, 110, 106, 109, 116, 119, 107, 106, 105, 108. Is this group significantly different from the first group?

*Answer*

*a.* If the group were larger we could apply a test based on the Normal distribution. Here we must apply the t-test.

The mean and standard deviation for the group are, respectively, 117·67 and 7·44 [calculated as in earlier chapters].

$$t = \frac{(117{\cdot}67 - 115)\sqrt{12-1}}{7{\cdot}44}$$

$$= 1{\cdot}19$$

The tables indicate that, for 12—1=11 degrees of freedom this value would occur about 25% of the time, so there is no significant difference. Thus the group may reasonably be supposed to come from a population with an average IQ of 115.

*b.* The mean and standard deviation of the second group are 110·3 and 4·86 respectively.

The difference in the means=117·67—110·3=7·37.

The estimate of the combined standard deviation is $\tilde{\sigma} =$

$$\sqrt{\frac{12 \times 7{\cdot}44^2 + 10 \times 486^2}{12 + 10 - 2}}$$

=6·71, giving the standard error of the difference between the means as $6{\cdot}71\sqrt{(\frac{1}{12} + \frac{1}{10})}$.

Dividing this into the difference in the means (7·37) results in a value of t of 2·57 for 12+10—2=20 degrees of freedom.

If we apply a two-tailed test at the 5% level we find that this value of 2·57 is greater than the tabulated value of 2·09, so there is a significant difference. At the 1% level the value would have to exceed 2·85, so it is not highly significant.

We can also test the significance of the difference between the variability shown in the two groups, using the F-test.

The variance of the first group is 60·42 and of the second 23·61, so that $F = \frac{60{\cdot}42}{23{\cdot}61} = 2{\cdot}62$. For degrees of freedom of 11 and 9 respectively the tables show that a value of 3·1 is significant at the 5% level. Since 2·62 is less than this we conclude that the variabilities shown within the two groups are not significantly different.

## 20.8.2 FURTHER EXAMPLE

*Question*

*a.* A sample of 10 items yields a correlation coefficient of 0·7. Is it reasonable to suppose that such a value would arise from a population where the coefficient is 0·85?

*b.* Is the value of 0·7 itself significant?

*c.* Another sample of 12 items shows a coefficient of 0·9. Is this likely to be from the same population as the first sample?

*Answer*

*a.* We convert 0·7 and 0·85 into corresponding Z values. Population Z=1·256 and sample Z=0·867. The difference is 0·389. The standard deviation Z is $\frac{1}{\sqrt{10-3}}$=0·378. The final value of Z for comparison with the normal tables is $\frac{0{\cdot}389}{0{\cdot}378}$=1·029.

Applying the one-tail test at the 5% level we should require a value of at least 1·64 to be significant. We conclude that the sample might well arise in the assumed population.

*b.* We wish to state whether r differs significantly from zero. With the small sample the t-test is appropriate, and $t=\frac{0{\cdot}7\sqrt{10-2}}{\sqrt{1-0{\cdot}7^2}}=2{\cdot}772$. For 8 degrees of freedom we require a value of 1·86 at the 5% level, using the one-tail test. Thus we conclude that the correlation is significant.

*c.* Here we return to the Z-test. Z for the sample of 10 (r=0·7) is 0·867 and for the sample of 12 (r=0·9) is 1·472. The difference is 1·472−0·867=0·605. The standard error is $\sqrt{\frac{1}{10-3}+\frac{1}{12-3}}$=0·504. Z is thus $\frac{0{\cdot}605}{0{\cdot}504}$=1·2, which is less than the 1·64 required at the 5% level. The two samples may thus well be from the same population.

## 20.9 PRACTICE QUESTIONS

1. Subtract 16 from each of the entries in table 20.2 and analyse the variance of these figures.

2. For the data of table 6.1 analyse the total variance into variance within the performance of each salesman and between the means of the salesmen.

3. For the data of table 10.1 perform analyses of variance:
   a. For the first 5 samples.
   *b.* For the last 7 samples.
   *c.* For all the samples.
   (You should reduce your working by coding the data.)

4. In certain market survey work a research organisation divides the population into 5 categories which are supposed to occur in given proportions. An investigation involving 20,000 people chosen randomly shows these categories to occur with the frequencies shown below. Test the agreement of these two sets of figures and comment on the practical implications.

| Category | A | B | C | D | E |
|---|---|---|---|---|---|
| Supposed % | 12 | 25 | 20 | 30 | 13 |
| From 20,000 | 2200 | 6040 | 4629 | 5867 | 1264 |

5. Of 50 people who read a particular paper 32 also purchase a particular product, while of 80 people who do not read the paper 25 buy this product.

   *a.* Put this information in the form of a contingency table.

   *b.* Does it suggest an association between readership and purchase?

6. A sample of 10 items has the following values: 37, 42, 38, 35, 43, 45, 40, 41, 39, 37. Does the mean of these differ significantly *a.* from 39; *b.* from 36?

7. A sample of 15 items has a mean of 13·2 and a standard deviation of 0·9. Is it likely to have arisen from a population with a mean of 13·5?

8. Two independent samples containing 6 and 9 items respectively yield the following values:

   Sample A 12 8 11 13 7 9
   Sample B 8 13 12 10 8 8 9 13 12.

   Is there a significant difference between the means of the samples?

9. A sample of 16 pairs of observations from a normal population yields a coefficient of correlation of 0·4. Test the significance of this.

10. Is there a significant difference between the variances of the two samples in qu. 8?

11. Assuming that the standard deviation for all items of the bonus data of table 2.1 is 0·8 apply the F-test to find if the variability of the column averages is significant. Make use of table 20.1.

12. A population is known to have a correlation coefficient of 0·45. A sample of 50 pairs of observations yields a coefficient of 0·61. Is the sample significantly different from the population?

13. Assuming that the samples of qu. 9 and qu. 12 relate to the same type of data are the samples significantly different?

14. Apply non-parametric tests to the data of questions 6 and 4.

# 21

# Some Mathematical Points

## 21.1 REVISION OF LOGARITHMS IN CALCULATIONS

Logarithms, as aids to calculation, enable us to multiply numbers by a process of addition.

*Common logarithms* use a *base* of 10. Logarithm tables show the power to which this must be raised to be equivalent to each number concerned. If we add these logarithms the result is the power of 10 equivalent to the product of the numbers. Thus $100 \times 1{,}000 = 10^2 \times 10^3 = 10^5 = 100{,}000$. Here the addition of 2 and 3 results in 5 which is the logarithm of 100,000. *Anti-logarithm* tables enable the conversion back from logarithmic form to be made.

To simplify the production of tables each number is considered as being the product of two parts, a number from 1 and up to but not including 10, multiplied by a power of 10. Thus $19{\cdot}68 = 1{\cdot}968 \times 10^1$; $1968 = 1{\cdot}968 \times 10^3$; $0{\cdot}01968 = \dfrac{1{\cdot}968}{10^2} = 1{\cdot}968 \times 10^{-2}$.

In the tables we should look up 1968 in each of these cases. With 4-figure logarithms we look down the left-hand column until the row having 19 is reached. We then read across to the column headed 6 and find the number 2923. The end 9 columns of mean differences are used to deal with the 4th figure. Under the column headed 8 we find 17, which is added to 2923 to give 2940.

This is termed the *mantissa* and is always regarded as following the decimal point in the logarithm. For the 3 numbers it is 0·2940 in each case.

The power of 10 which is to be used as a multiplier is termed the *characteristic* of the logarithm and is placed to the left of the mantissa. Thus, the log of 19·68 is 1·2940, that of 1968 is 3·2940, while that of 0·01968 is $\bar{2}{\cdot}2940$. Notice that a negative characteristic is shown with a bar above it, read *bar 2 point 2940*. This is because the log is really $-2+0{\cdot}2940$ and not $-2{\cdot}2940$.

A simple rule for the characteristic is that it is 1 less than the number of places to the left of the point in numbers greater than 1. For numbers less than 1 it is 1 more than the number of zeros immediately following the point, and is negative.

When using logarithms it is advisable to place the numbers and their logs neatly in two columns. Thus, to multiply $3{\cdot}67 \times 48{\cdot}5 \times 0{\cdot}64$ we have:

| No. | Log. |
|---|---|
| 367 | 2·5647 |
| 48·5 | 1·6857 |
| 0·64 | $\bar{1}{\cdot}8062$ |
| | 4·0566 |

In totalling the logs we add the mantissa part in the normal way. If this results in any units they are added into the characteristic column, taking account of any negatives (bar values) which may be present. Thus we have 2·0566 for the addition of the mantissas. The 2 is added to the $2+1-1$ in the characteristic column, resulting in 4·0566.

In reading the antilogarithm tables we look up only the mantissa part of the log. In this case it is 0·0566. The first two figures following the point (05) give the row. Moving across to the main column headed 6 we find 1138. To this is added the value under the mean difference column 6, to give 1140. The characteristic (4) tells us there must be $4+1=5$ places to the left of the point, so we have the answer 11,400.

A negative characteristic implies that the answer is less than 1, with 1 less zero than the characteristic. Thus $\bar{4}$·0566 would be equivalent to 0·000114.

*Division* implies subtraction of logs. The calculation for 367÷48·5 is shown at (a) below. That for 48·5÷367 is shown at (b). Here, in the characteristic column, we are taking 2 from 1 (i.e. $1-2$) leaving $\bar{1}$. If we have 48·5÷0·64 (shown at (c) ) the mantissa subtraction gives 0·8795 with 1 'borrowed'. The characteristic is thus $1-(-1)-1$ (borrowed) $=1+1-1=1$.

| (a) No. | Log | (b) No. | Log | (c) No. | Log |
|---|---|---|---|---|---|
| 367 | 2·5647 | 48·5 | 1·6857 | 48·5 | 1·6857 |
| 48·5 | 1·6857 | 367 | 2·5647 | 0·64 | $\bar{1}$·8062 |
| | 0·8790 | | $\bar{1}$·1210 | | 1·8795 |
| antilog | 7·568 | antilog | 0·1321 | antilog | 75·77 |

*Powers* of numbers can be obtained by multiplying the logarithm by the power required, and *roots* by dividing.

The calculation for $3{\cdot}67^3$ involves $0{\cdot}5647\times3=1{\cdot}6941$, of which the antilog is 49·44.

$0{\cdot}64^4$ calls for $\bar{1}{\cdot}8062\times4=\bar{1}{\cdot}2248$, of which the antilog is 0·1678. In the multiplication we have $0{\cdot}8062\times4=3{\cdot}2248$ for the mantissa, the 3 being added to $4\times(-1)=-4$ for the characteristic, resulting in $3-4=\bar{1}$.

With roots, say $\sqrt[3]{0{\cdot}64}$, we have $\bar{1}{\cdot}8062\div3=\bar{1}{\cdot}9354$, of which the antilog is 0·8618. In cases of division such as this, where a bar characteristic is involved, we make up the characteristic so that it will divide exactly, and then add the reverse amount to the mantissa. We have $\bar{1}$ made up to $\bar{3}$ so that it divides to give $\bar{1}$. This involved the addition of $-2$ to the log so we adjust the mantissa to 2·8062, which divides to give 0·9354. With the characteristic this becomes $\bar{1}$·9354.

*Squares* and *square roots* are so important in statistics that special tables are available which we look up in the same way as logarithm tables. Care must be taken with roots since the position of the decimal point is vital. The roots of 160, 1·6 and 0·016 are 12·6, 1·26 and 0·126 respectively, while the roots of 1,600, 16 and 0·16 are 40, 4, and 0·4 respectively. The root tables are divided into the roots of numbers from 1 to 10 and those from 10 to 100 to allow for this. If we have an even number of figures to the left of the point, or an even number of zeros (or none) immediately following the point we should use the 10 to 100 part of the table. The position of the point in the answer can be obtained by considering that every 2 moves left in the original number implies division by 100 of which the root is 10 so we must move 1 place to the right in the answer (and the reverse for numbers less than 1).

In the specimen logarithm calculations all the figures of the antilog have been retained in the answer. The last figure of the result cannot be regarded as accurate, however, and we are only justified in giving the results to 3 *significant figures*, which would result in 49·4, 0·168, and 0·862 for the calculations of the powers. This implies we retain the first three figures of the four obtained from the antilog tables. It is not the same as working to three *decimal places*, where three figures will always follow the decimal point.

## 21.2 THE SLIDE RULE

The operation of this very handy, cheap and simple calculating device should be understood by all who have to deal with statistical calculations. Multiplication, division, squares and square roots, and many other calculation can be performed rapidly. Students should gain experience with their own slide rule so that they can rely on it for examination and more general use.

The principle is easy to grasp. Diagram 21.1 attempts to show the basis of slide rule design.

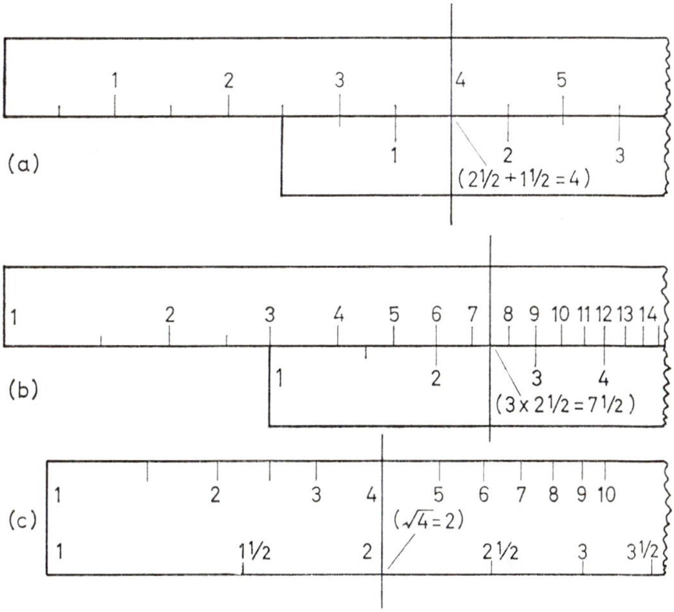


Diagram 21.1. *Slide rule principles*

The upper part shows how we may add numbers by using two similarly graduated scales, one of which is moved along the other. Here we add $2\frac{1}{2}$ and $1\frac{1}{2}$ by moving the zero of one scale to coincide with $2\frac{1}{2}$ on the other. Opposite $1\frac{1}{2}$ on the lower scale we find 4 on the upper. Subtraction is the reverse of this. Try it with two rulers. It is not a very practical way of adding, but if the numbers are logarithms the adding amounts to multiplication, which is very useful.

In the centre diagram we see two scales where the numbers are unevenly spaced. In fact, they are spaced at distances proportional to their logarithms. Thus, since the log of 1 is zero, 1 appears at the start of the scale. Numbers 2, 4, and 8 are equally spaced, since their logs are 0·301, 0·602 and 0·903 respectively. The operation of multiplying $3\times2\frac{1}{2}$ is shown, by a process exactly similar to that just explained for addition of ordinary numbers. The reverse process provides division.

The lower part of the diagram shows that, if one scale has its numbers, as logarithms, spread over twice the distance of another scale, aligned with it at zero, then we can read off squares and square roots.

Most slide rules have scales from 1 to 10 running the whole length, and other scales, less widely spaced, so that two of these fit into this length. This is necessary with roots, since these run in two series; thus, the square roots of 4,400, 40,000 etc. are 2, 20, 200 etc., while the roots of 40, 4,000, 400,000 etc. are 6·325, 63·25, 632·5 etc.

In using the slide rule, except in the case of square roots, it is best to ignore the decimal points, these being inserted in the answer from an inspection of the original data.

## 21.3 SERIES

A mathematical *series* consists of a sequence of numbers (or symbols) called *terms* which are related by a definite law. Knowing one term we can obtain the rest.

Data which form a series are usually more easily handled than figures which have no underlying pattern. Much of statistical method is concerned with finding series which approximately fit data.

### 21.3.1 ARITHMETIC SERIES

The simplest series is, perhaps, the *arithmetic* progression (AP) in which a fixed amount called the *common difference* is added to each number, in turn, to arrive at the following number.

The numbers 1, 3, 5, 7 are in AP with a common difference of 2; the numbers 10, 6, 2, −2, −6 have a common difference of −4.

Suppose that there are n terms, of which the first is a and the last is L, the common difference being d and the sum of the terms S.

The nth term will be $a + (n-1)d$ and the average of the terms will be $\frac{a+L}{2}$ (the average of the first and last terms since all terms are equally spaced). The sum of all the terms will be given by $S=\frac{n(a+L)}{2}$ (since this is the number of terms multiplied by their average).

A good commercial example of AP is the growth of a sum of money at *simple interest*. The amounts each year are in AP with a common difference= principal$\times\frac{R}{100}$, where R is the yearly interest rate per cent.

Thus, £200 at 5% gives the terms 200, 210, 220, 230 etc.

We have an example of a negative common difference in the depreciation of a capital asset by a fixed amount each year (*straight-line depreciation*).

For an asset costing £200 the depreciated values might be 190, 180, 170, 160, 150 etc.

### 21.3.2 GEOMETRIC SERIES

Here each term is some multiple (called the *common ratio*) of the previous term.

The numbers 1, 3, 9, 27 are in geometric progression (GP) with 3 as the common ratio; the numbers 32, 8, 2, $\frac{1}{2}$, $\frac{1}{8}$ are in GP with $\frac{1}{4}$ as the common ratio.

If we denote the ratio by r then the terms are a, ar, $ar^2$, $ar^3$ . . . . $ar^{n-1}$.

*Compound interest* offers a good example of GP. The first term is the principal (denoted by P) and the common ratio is $\left(1+\frac{R}{100}\right)$, where R is the rate of interest. The amount (A) at the end of N years is given by $A=P\left(1+\frac{R}{100}\right)^N$.

Note that the principal is considered to be invested at the start of the first year (or at the end of year 0), so that the Nth year is the (n+1)th term.

If we reduce the amount each year by a common ratio $\left(1-\frac{R}{100}\right)$ we have, as an example, the asset value under the *diminishing balance* type of depreciation. The depreciation is a fixed percentage of the value at the end of each year.

For comparison with the simple interest example the amounts of £200 at 5% compound interest form a series (200), 210, 220·5, 231·525 for the first 3 years; the diminishing balances at 5% result in 190, 180·5, 171·475 for the first 3 years.

The *present value* (PV) of an amount (A) in N years is the principal which would have to be invested now to amount to A in N years. From $A=P\left(1+\frac{R}{100}\right)^N$ we see that $P=\frac{A}{\left(1+\frac{R}{100}\right)^N}$.

Compound interest and present value tables are usually standardised to show the amount of £1 in N years and the present value of £1 in N years. We multiply the value in the tables by our given value of P or A, as the case may be.

*Discounted cash flow* uses PVs. The cash flowing each year from an investment (in equipment etc.) is multiplied by the PV for that year. The sum of the products gives the present value of these future benefits so that they can be compared with the cost of the investment. Several projects may be evaluated in this way and the most satisfactory one selected. In this case we are assuming that the rate of interest is known. This is sometimes called an *external* rate of interest and the result of the calculation the *net present value*. An alternative is to calculate the rate of interest (termed the *yield*) that would make the present value of the returns from the investment equal to the sum invested. This latter involves trial and error or interpolation or, preferably, a computer programmed to give the required yield.

The sum of a GP is given by $S = a + ar + ar^2 + ar^3 + \dots + ar^{n-1}$. If we multiply this throughout by r we have $rS = ar + ar^2 + ar^3 + \dots + \dots ar^n$. Subtraction of the original series from this gives $S(r-1) = ar^n - a$, from which we see that $S = \frac{a(r^n-1)}{r-1} = \frac{a(1-r^n)}{1-r}$. If r is less than 1 the term $r^n$ will become negligible if a large number of terms is involved. In this case the *sum to infinity* $= \frac{a}{1-r}$.

An *annuity* implies that a fixed payment is made each year. If p is the yearly payment and R is the rate of compound interest then the value of this stream of payments at the end of N years will be $S = p\left(1+\frac{R}{100}\right) + p\left(1+\frac{R}{100}\right)^2 + \dots + p\left(1+\frac{R}{100}\right)^N$. The first term represents the value of the last £p invested a year ago, the second the value of the previous £p of 2 years ago, while the last term represents the value of the first £p invested. Using the formula for the sum of a GP we see that the value of the annuity amounts to $\frac{p\left(\left(1+\frac{R}{100}\right)^N - 1\right)}{\left(1+\frac{R}{100}\right) - 1}$. Notice that the common ratio (r) is $1+\frac{R}{100}$, not R, the rate of interest.

### 21.3.3 BINOMIAL SERIES

This is the series obtained by expanding $(a+b)^n$. Check by multiplication ( $(a+b)^2 = (a+b)(a+b)$, etc) that, for the first few values of n, this gives:

$$(a+b)^1 = a+b$$
$$(a+b)^2 = a^2+2ab+b^2$$
$$(a+b)^3 = a^3+3a^2b+3ab^2+b^3$$
$$(a+b)^4 = a^4+4a^3b+6a^2b^2+4ab^3+b^4.$$

A general rule which gives the required terms for positive integer values of n is:

$$(a+b)^n = a^n + \frac{n}{1!}a^{n-1}b + \frac{n(n-1)}{2!}a^{n-2}b^2 + \frac{n(n-1)(n-2)}{3!}a^{n-3}b^3 + \dots\dots + b^n.$$

*Factorial notation* n! implies $1 \times 2 \times 3 \times 4 \times \dots \times n$; for example $4! = 1 \times 2 \times 3 \times 4$.

In statistics we often use p and q instead of a and b. These denote the probabilities of something happening and of it not happening, respectively, and must total 1.

In mathematics the expansion of $(1+x)^n$ receives much attention. It is a special case of the above and

$$(1+x)^n = 1 + nx + \frac{n(n-1)}{2!}x^2 + \frac{n(n-1)(n-2)}{3!}x^3 + \dots + x^n,$$

which is true for all values of x.

If n is not a positive integer the expansion holds good only if x is between +1 and −1.

## 21.3.4 EXPONENTIAL SERIES AND e

The constant e has the value 2·71828. It features in much statistical work, examples being the normal distribution, the Poisson series, and growth curves. It is also used as the base in *natural* logarithms.

This valuable constant can be regarded as the sum of a series of rapidly diminishing numbers:

$$e=1+\frac{1}{1!}+\frac{1}{2!}+\frac{1}{3!}+\ldots\ldots$$

Calculation of the first few terms soon gives a value close to that quoted above.

An *exponent* is the *power* of a number. An important series with x as the power is that for $y=e^x$. It can be shown that

$$e^x=1+\frac{x}{1!}+\frac{x^2}{2!}+\frac{x^3}{3!}+\ldots\ldots$$

It will be seen that, when $x=1$, this becomes the series used to calculate e above.

The *Poisson distribution*, used as an approximation to the binomial distribution when we are considering small probabilities and large samples is:

$e^{-m}\left(1+\frac{m}{1!}+\frac{m^2}{2!}+\frac{m^3}{3!}+\ldots\ldots\right)$, where m is the expectation of the event, i.e. number of items $\times$ probability.

This is the equivalent of $e^{-m}\times e^m=e^0=1$, which must be the total of all the probabilities involved.

One of the most interesting characteristics of $y=e^x$ is that the rate of change of y is proportional to y itself. This follows from the fact that the derivative of $e^x$ (see section 21.5) is itself $e^x$. If $y=e^{kx}$ then the derivative is $\frac{dy}{dx}=ke^{kx}$.

## 21.4.1 SIMPLE EQUATIONS

Suppose we think of a number, multiply it by 3, and then add 7. The result is 19. What is the number?

Let x equal the number. Then the calculation is equivalent to $3x+7=19$. Subtract 7 from each side, resulting in $3x=12$. Divide each side by 3, resulting in $x=4$, which is the required number.

Equations imply that the left and right hand sides are equal, so we may add (subtract) the same amount to (from) each, or multiply (divide) each by the same amount.

## 21.4.2 SIMULTANEOUS EQUATIONS

These are equations with two or more unknowns. We require as many equations as there are unknowns.

If the equations are

$$5x+2y=16 \quad \ldots\ldots\ldots\ldots \quad (1)$$
$$3x+4y=18 \quad \ldots\ldots\ldots\ldots \quad (2)$$

then we wish to manipulate these so that we have a single equation involving either x or y.

Multiplying (1) by 2 we have

$$10x+4y=32 \quad \ldots\ldots\ldots\ldots\ldots\ldots\ldots\ldots\ldots \quad (3).$$

Subtracting (2) from (3) gives

$$7x=14$$

so that $x=2$.

Substitution in, say, (1) gives

$$10+2y=16$$
$$2y=6$$
$$y=3.$$

Draw graphs of the two straight lines and check that the above solution gives the point of intersection.

An easy way of drawing such lines is to let $x=0$ and find the value of y,($0+2y=16$, so $y=8$, for equation (1) ) and then let $y=0$ and find the value of x, ($5x+0=16$, so $x=3\frac{1}{5}$). This gives a point on each axis, which is sufficient to locate the line.

### 21.4.3 QUADRATICS

A quadratic involves squares, an example being

$$x^2+4x-45=0.$$

If it can be factorised the solution is easy. In this case we have

$$(x+9)(x-5)=0$$

so that, either $x+9=0$, i.e. $x=-9$
or $x-5=0$, i.e. $x=5$.

Both of these satisfy the equation.

If we plot $y=x^2+4x-45$ then the curve is U-shaped and will be found to cut the x axis (where $y=0$) when $x=-9$ and $x=5$.

In more difficult cases it is more satisfactory to use a formula.

Suppose that the quadratic is

$$ax^2+bx+c=0,$$

where a, b and c are constants in a particular case.

Divide throughout by a, giving

$$x^2+\frac{bx}{a}+\frac{c}{a}=0.$$

Subtract $\frac{c}{a}$ from each side to give

$$x^2+\frac{bx}{a}=-\frac{c}{a}.$$

Add $\left(\frac{b}{2a}\right)^2$ to each side, giving

$$x^2+\frac{bx}{a}+\left(\frac{b}{2a}\right)^2=-\frac{c}{a}+\left(\frac{b}{2a}\right)^2=\frac{-4ac+b^2}{4a^2}.$$

The left hand side factorises to $\left(x+\frac{b}{2a}\right)^2$.

If we take the square root of each side (remembering that there is a + and − root) we have

$$x+\frac{b}{2a}=\pm\frac{\sqrt{b^2-4ac}}{2a}.$$

so that $x=\dfrac{-b\pm\sqrt{b^2-4ac}}{2a}$.

Applying this to the example used previously, where a=1, b=4 and c=−45, we have $x=\dfrac{-4\pm\sqrt{16-4\times1\times-45}}{2\times1}=\dfrac{-4\pm14}{2}=-9$ or 5.

If $b^2$ is less than 4ac the square root of a negative number is involved and the roots are said to be imaginary.

## 21.5 SIMPLE CALCULUS, DERIVATIVES, INTEGRALS

*Differential calculus* is of practical use for the manipulation of algebraic equations, when we are concerned with the rate of change of one factor compared with another.

We have seen in chapter 11 that a straight line takes the form $y=A+Bx$, and that B represents the increase in y per unit increase in x, or the gradient of the line. With a straight line the rate of increase is constant throughout the whole of its length (diagram 11.1).

With a curve the rate of increase differs for different values of x. At any point on the curve it can be estimated by drawing a tangent to the curve and finding its gradient.

The calculus offers a simple solution to this problem without the need for drawing the graph. The process is called *differentiation.* In doing this terms of the type $x^n$ become $nx^{n-1}$ (called the *derivative*), on the right hand side of the equation, while we replace y by $\frac{dy}{dx}$ (pronounced dee-y by dee-x) on the left, indicating that the effect is to show the small increase in y resulting from a small increase in x.

If the equation is $y=7+3x+8x^2$ we have $\frac{dy}{dx}=3+16x$. Note that constant terms disappear since they do not affect the gradient.

Substitution of any desired value of x in this derivative gives the gradient at this point.

When x=0 the gradient is 3+0=3; when x=3 the gradient is 3+48=51.

A very important point occurs where the gradient is zero, since this implies that the curve is at a maximum or minimum.

If we wish to find the point at which the curve $y=7+3x+8x^2$ is in this condition we let $\frac{dy}{dx}=0$ and solve for x. This gives $3+16x=0$ from which $x=-\frac{3}{16}$.

Is this a maximum or minimum point? Suppose we differentiate again giving $\frac{d^2y}{dx^2}$ (dee-two-y by dee-x-squared)=16. This gives the rate at which the gradient is increasing with x. If it is positive (when we substitute our value of x) the gradient is tending to become steeper, indicating we are passing through a minimum point. If it is negative the reverse applies and we have a maximum. In this case the differential has given a constant +16 so the point must be a minimum.

The reader who is unfamiliar with calculus should draw this curve and check the gradients at different points.

The reason for the rule for differentiation given above can be seen if we consider an equation of the type $y=x^n$. When $x=a$ the value of y is $a^n$. If x increases by a very small amount b then it has become $(a+b)$ and $y=(a+b)^n$. The binomial expansion of this gives $a^n+na^{n-1}b$, since we can ignore further terms as they involve higher powers of the very small amount b and are negligible. The increase in y is thus $(a^n+na^{n-1}b)-a^n=na^{n-1}b$. This has occurred while x has increased by b. The gradient is thus $\frac{na^{n-1}b}{b}=na^{n-1}$.

The rule works for negative powers of x. If $y=\frac{1}{x^2}=x^{-2}$ then $\frac{dy}{dx}=-2x^{-3}=-\frac{2}{x^3}$.

The opposite process to differentiation is called *integration*. Terms of the type $x^n$ integrate to become $\frac{x^{n+1}}{n+1}$. Using this we can convert from an equation showing the gradients to the basic equation to which these gradients apply. The constant term, which fixes the height of this graph, requires the substitution of actual points to establish it.

Suppose we integrate the equation $y=3x+1$. This is written $\int y\,dx$ (integral y-dee-x) or $\int(3x+1)dx$. Using the rule just given we obtain $\frac{3x^2}{2}+x+K$, where K is a constant. If we know that the value of this expression is 12 when $x=2$ then we have $12=\frac{3}{2}\times 2^2+2+K$, so that $K=4$ and the integral is $\frac{3x^2}{2}+x+4$.

It is most useful to regard the integral as providing the sum of all the small increments between two points on the x axis. This sum is equal to the area under the curve between these points.

Suppose that we wish to calculate the area under the curve $y=7+3x+8x^2$ between the points where $x=0$ and $x=2$. We require $\int_0^2(7+3x+8x^2)dx$. The integral is $\left[K+7x+\frac{3x^2}{2}+\frac{8x^3}{3}\right]_0^2$ which, placing $x=2$ and $x=0$, gives $[K+14+6+21\tfrac{1}{3}]-[K+0+0+0]=41\tfrac{1}{3}$ as the area. In such cases it is not necessary to find K, since it disappears.

The integral of $\frac{1}{x}$ presents some problems. If we apply the general rule we have $\int x^{-1}$ which seems to result in $\frac{x^0}{0}$. In fact the integral is $\log_e x$, a case where natural logarithms are important.

## 21.6 BINARY NUMBERS

Reference to binary notation was made in the section on digital computers. All numbers are built up from powers of the base 2, the position, marked by 1, indicating which of these is involved. There are only two characters, either 1, which indicates that the position is filled, or 0, which indicates that it is empty.

Three binary numbers and their decimal equivalents are shown below.

| power | $2^5$ | $2^4$ | $2^3$ | $2^2$ | $2^1$ | $2^0$ | |
|---|---|---|---|---|---|---|---|
| value | 32 | 16 | 8 | 4 | 2 | 1 | Decimal equivalent |
| | | | | 1 | 0 | 1 | 5 |
| | | 1 | 1 | 0 | 1 | 0 | 26 |
| | | | 1 | 0 | 0 | 0 | 8 |

*Conversion* from decimal to binary form is by successive division by 2, the remainders and the last 1 giving the binary number, in reverse order.

Thus

| | | |
|---|---|---|
| 2)26 | and | 2)8 |
| 2)13 rem. 0 | | 2)4 rem. 0 |
| 2) 6 rem. 1 | | 2)2 rem. 0 |
| 2) 3 rem. 0 | | 1 rem. 0 |
| 1 rem. 1 | | |

Conversion from binary to decimal is best done by adding the values of the different positions, as indicated in the opening table.

*Addition* of binary numbers is easy if we remember that the highest value in any position is 1. Every time we have $1+1$ we must add 1 to the next column to the left.

Thus $26+8$ involves

| | | | | | | |
|---|---|---|---|---|---|---|
| 26= | | 1 | 1 | 0 | 1 | 0 |
| +8= | | | 1 | 0 | 0 | 0 |
| 34= | 1 | 0 | 0 | 0 | 1 | 0. |

Subtraction is similar, remembering that $0-1$ involves borrowing 1 from the next column on the left.

The calculation for $26-5$ is

| | | | | | |
|---|---|---|---|---|---|
| 26= | 1 | 1 | 0 | 1 | 0 |
| −5= | | | 1 | 0 | 1 |
| 21= | 1 | 0 | 1 | 0 | 1 |

*Multiplication* involves shifts of the multiplicand to the left according to the position of the 1's in the multiplier.

The multiplication of $26 \times 5$ is equivalent to

$$
\begin{array}{rrl}
26\times 1 = & 26 = & \phantom{1\,1\,\,}1\;1\;0\;1\;0 \\
26\times 4 = & 104 = & \phantom{1\,}1\;1\;0\;1\;0 \\
\hline
& 130 = & 1\;0\;0\;0\;0\;0\;1\;0
\end{array}
$$

There are no multiplication tables to learn as everything depends on three additions, $0+0=0$, $0+1=1$ and $1+1=10$. This is the reason for the choice of this system for computers.

Division involves the reverse procedure.

We place the divisor as far as we can to the left, so that it does not exceed the number being divided, and record the number of shifts to the left. We then subtract and repeat, again counting the number of shifts. The total of the shifts gives the decimal equivalent of the answer.

In the case of $130 \div 26$ we have

$$
\begin{array}{rll}
130 = & 1\;0\;0\;0\;0\;0\;1\;0 & \\
(26) & \phantom{1\,}1\;1\;0\;1\;0 & \text{divisor moved 2 positions} = 4 \\
\hline
 & 0\;0\;0\;1\;1\;0\;1\;0 & \\
(26) & \phantom{0\,0\,0\,}1\;1\;0\;1\;0 & \text{divisor moved 0 positions} = 1 \\
\hline
\text{remainder} = & \phantom{0\,0\,0\,}0\;0\;0\;0\;0 & \text{quotient} = 5
\end{array}
$$

If there is a remainder the process can be continued into positions to the right of a binary point, the values being $\frac{1}{2}$, $\frac{1}{4}$, $\frac{1}{8}$ etc.

## 21.7 STATISTICAL ALGEBRA

Many of the proofs for statistical formulae involve handling summations where, for example, $\Sigma x$ = sum of the x values, $\Sigma(x-A)$ = sum of the differences of the x values from A.

This should cause no difficulty if it is remembered that:

a. The summation sign and the values to which it relates can be treated as a single variable.

b. Any brackets following the sign can be expanded according to the normal rules of algebra.

c. The position of squared signs etc. is vital. Thus $\Sigma(x-A)^2$ is quite different from $(\Sigma(x-A))^2$. The former implies that we first square each deviation from A and then sum these squares. The latter implies that we add the deviations and then square the total.

An example of expansion is $\Sigma(x-A)=\Sigma x-\Sigma A$. If there are n items then A (a constant) will appear n times so that $\Sigma A=nA$ and $\Sigma(x-A)=\Sigma x-nA$.

Again $\Sigma(x-A)^2=\Sigma(x^2-2Ax+A^2)=\Sigma x^2-2A\Sigma x+nA^2$.

If A is an assumed mean then $(x-A)$ is a deviation from this and $\Sigma(x-A)$ is the sum of such deviations. We know that the mean,

$$\bar{x}=\frac{\Sigma x}{n}=\frac{\Sigma(x-A+A)}{n}=\frac{\Sigma(x-A)}{n}+\frac{\Sigma A}{n}=\frac{\Sigma(x-A)}{n}+A.$$

This implies that we add to the assumed mean (A) the sum of the differences of the data from the assumed mean, divided by the number of items.

The variance is $\frac{\Sigma(x-\bar{x})^2}{n}=\frac{\Sigma x^2}{n}-\frac{2\bar{x}\Sigma x}{n}+\frac{n\bar{x}^2}{n}$

$$=\frac{\Sigma x^2}{n}-2\bar{x}^2+\bar{x}^2$$

$=\frac{\Sigma x^2}{n}-\bar{x}^2$, which is the formula for the variance in terms of the squared data.

Try yourself to prove the formula for the standard deviation working from an assumed mean.

The expansions of $\Sigma f(d+1)=\Sigma fd+n$ and $\Sigma f(d+1)^2=\Sigma fd^2+2\Sigma fd+n$ are the basis of *Charlier's checks* in computing the mean and standard deviation. In these cases we wish to check the totals $\Sigma fd$ and $\Sigma fd^2$. This is done by adding 1 to each d value and calculating $\Sigma f(d+1)$ and $\Sigma f(d+1)^2$, in addition to the normal calculation. We can then check if the identities above apply with the figures we are using.

# Summary Tables

Table 1. *Ordinates and areas of the normal curve*

| z | Ordinate | Area from z=0 up to z |
|---|---|---|
| 0·0 | 0·3989 | 0·0000 |
| 0·5 | 0·3521 | 0·1915 |
| 1·0 | 0·2420 | 0·3413 |
| 1·5 | 0·1295 | 0·4332 |
| 2·0 | 0·0540 | 0·4772 |
| 2·5 | 0·0175 | 0·4938 |
| 3·0 | 0·0044 | 0·4987 |
| 3·5 | 0·0009 | 0·4998 |

The distribution is symmetrical about z=0.

$$z=\frac{\text{deviation from mean}}{\text{standard deviation}}.$$

For the total area to the left of z add 0·5 to the areas in the final column.

For the upper (and lower) tail values subtract the areas from 0·5.

Table 2. *Percentage points of the normal curve*

| % Cut Off | Value of z at which this occurs. Single-Tail | Two-Tail |
|---|---|---|
| 20 | 0·842 | 1·282 |
| 10 | 1·282 | 1·645 |
| 5 | 1·645 | 1·960 |
| 1 | 2·326 | 2·576 |

Table 3. *Values of* $e^{-m}$

| m | $e^{-m}$ | m | $e^{-m}$ |
|---|---|---|---|
| 0·0 | 1·0000 | 1 | 0·3679 |
| 0·1 | 0·9048 | 2 | 0·1353 |
| 0·2 | 0·8187 | 3 | 0·0498 |
| 0·3 | 0·7408 | 4 | 0·0183 |
| 0·4 | 0·6703 | 5 | $0{\cdot}0^{2}674$ |
| 0·5 | 0·6065 | 6 | $0{\cdot}0^{2}248$ |
| 0·6 | 0·5488 | 7 | $0{\cdot}0^{3}912$ |
| 0·7 | 0·4966 | 8 | $0{\cdot}0^{3}335$ |
| 0·8 | 0·4493 | 9 | $0{\cdot}0^{3}123$ |
| 0·9 | 0·4066 | 10 | $0{\cdot}0^{4}454$ |

Other values may be calculated using the fact that $e^{a+b}=e^{a}\times e^{b}$; thus $e^{-4\cdot5}=e^{-4}\times e^{-0\cdot5}=\cdot0183\times\cdot6065=\cdot0111$.

Table 4. *Percentage points of the $\chi^2$- and t-distributions* (upper tail values)

| Degrees of freedom | $\chi^2$-Distribution | | | t-Distribution | | |
|---|---|---|---|---|---|---|
| ν | 95% | 5% | 1% | 5% | 2·5% | 1% |
| 1 | 0·004 | 3·8 | 6·6 | 6·31 | 12·71 | 31·82 |
| 2 | 0·10 | 6·0 | 9·2 | 2·92 | 4·30 | 6·97 |
| 3 | 0·35 | 7·8 | 11·3 | 2·35 | 3·18 | 4·54 |
| 4 | 0·71 | 9·5 | 13·3 | 2·13 | 2·78 | 3·75 |
| 5 | 1·15 | 11·1 | 15·1 | 2·02 | 2·57 | 3·37 |
| 6 | 1·64 | 12·6 | 16·8 | 1·94 | 2·45 | 3·14 |
| 7 | 2·17 | 14·1 | 18·5 | 1·90 | 2·37 | 3·00 |
| 8 | 2·73 | 15·5 | 20·1 | 1·86 | 2·31 | 2·90 |
| 9 | 3·33 | 16·9 | 21·7 | 1·83 | 2·26 | 2·82 |
| 10 | 3·94 | 18·3 | 23·2 | 1·81 | 2·23 | 2·76 |
| 11 | 4·57 | 19·7 | 24·7 | 1·80 | 2·20 | 2·72 |
| 12 | 5·23 | 21·0 | 26·2 | 1·78 | 2·18 | 2·68 |
| 13 | 5·89 | 22·4 | 27·7 | 1·77 | 2·16 | 2·65 |
| 14 | 6·57 | 23·7 | 29·1 | 1·76 | 2·15 | 2·62 |
| 15 | 7·26 | 25·0 | 30·6 | 1·75 | 2·13 | 2·60 |
| 20 | 10·85 | 31·4 | 37·6 | 1·73 | 2·09 | 2·53 |
| 25 | 14·61 | 37·7 | 44·3 | 1·71 | 2·06 | 2·49 |
| 30 | 18·49 | 43·8 | 50·9 | 1·70 | 2·04 | 2·46 |

For the t-distribution the single-tail values given are equivalent to 10%, 5% and 2% double-tail values.

Table 4 is taken from Tables III and IV of Fisher & Yates: *Statistical Tables for Biological, Agricultural and Medical Research*, published by Oliver & Boyd Ltd., Edinburgh, and by permission of the authors and publishers.

# Index

Sources of published statistics are not separately indexed. They will be found in chapter 18.